Lecture Notes in Bioinformatics 3388

Edited by S. Istrail, P. Pevzner, and M. Waterman

Editorial Board: A. Apostolico S. Brunak M. Gelfand
T. Lengauer S. Miyano G. Myers M.-F. Sagot D. Sankoff
R. Shamir T. Speed M. Vingron W. Wong

Subseries of Lecture Notes in Computer Science

W0079526

Jens Lagergren (Ed.)

Comparative Genomics

RECOMB 2004 International Workshop, RCG 2004
Bertinoro, Italy, October 16-19, 2004
Revised Selected Papers

 Springer

Series Editors

Sorin Istrail, Celera Genomics, Applied Biosystems, Rockville, MD, USA
Pavel Pevzner, University of California, San Diego, CA, USA
Michael Waterman, University of Southern California, Los Angeles, CA, USA

Volume Editor

Jens Lagergren
Stockholm Bioinformatics Center
Department of Numerical Analysis and Computer Science
KTH, 10691 Stockholm, Sweden
E-mail: jensl@nada.kth.se

Library of Congress Control Number: 2004118127

CR Subject Classification (1998): F.2, G.2, E.1, H.2.8, J.3

ISSN 0302-9743
ISBN 3-540-24455-7 Springer Berlin Heidelberg New York

This work is subject to copyright. All rights are reserved, whether the whole or part of the material is concerned, specifically the rights of translation, reprinting, re-use of illustrations, recitation, broadcasting, reproduction on microfilms or in any other way, and storage in data banks. Duplication of this publication or parts thereof is permitted only under the provisions of the German Copyright Law of September 9, 1965, in its current version, and permission for use must always be obtained from Springer. Violations are liable to prosecution under the German Copyright Law.

Springer is a part of Springer Science+Business Media

springeronline.com

© Springer-Verlag Berlin Heidelberg 2005
Printed in Germany

Typesetting: Camera-ready by author, data conversion by Olgun Computergrafik
Printed on acid-free paper   SPIN: 11380757   06/3142   5 4 3 2 1 0

Preface

The RECOMB Satellite Workshop on Comparative Genomics (RECOMB-CG) is a forum on all aspects and components of Comparative Genomics ranging from new quantitative discoveries about genome structures and processes to theorems on the complexity of computational problems inspired by genome comparison.

Due to the interdiciplinary nature of the workshop, papers could be submitted merely for presentation at the workshop or for presentation at the workshop and publication in the proceedings. Speakers presenting papers of the former category are listed under selected presentations.

The workshop was a great success scientifically as well as socially. I want to thank all participants, the members of the program committee as well as referees and, especially, the excellent invited speakers.

October 2004   Jens Lagergren

Invited Speakers

Mathieu Blanchette (McGill University, Canada)
Daniela Delneri (University of Manchester, UK)
Henrik Kaessmann (Université de Lausanne, Switzerland)
Martin Lercher (University of Bath, UK)
Bill Martin (Heinrich-Heine Universität, Germany)
Ben Raphael (UCSD, USA)
Marie-France Sagot (Université Lyon I, France)
Graziano Pesole (University of Milan, Italy)

Selected Presentations

Ali Bashir (UCSD, USA)
Inna Dubchak (Lawrence Livermore National Laboratory, USA)
Dannie Durand (Carnegie Mellon University, USA)
Ron Y. Pinter (Technion, Israel)

Organization

Scientific Organizing Committee

Jens Lagergren (SBC and KTH, Sweden)
Aoife McLysaght (University of Dublin, Ireland)
Nancy Moran (University of Arizona, USA)
Bernard Moret (University of New Mexico, USA)
David Sankoff (University of Ottawa, Canada)

Program Committee

Chair: Jens Lagergren (SBC and KTH, Sweden)
Lars Arvestad (SBC and KTH, Sweden)
Mathieu Blanchette (McGill University, Canada)
Dannie Durand (Carnegie Mellon University, USA)
Emmanuelle Lerat (University of Arizona, USA)
Aoife McLysaght (University of Dublin, Ireland)
Nancy Moran (University of Arizona, USA)
Pavel Pevzner (UCSD, USA)
David Sankoff (University of Ottawa, Canada)
Igor V. Sharakhov (Virginia Tech, USA)
Li-San Wang (University of Pennsylvania, USA)

Referees

G. Bourque	F. Ge	N. Song
L. Cui	R. Hoberman	L. Zhang
G.S. Ganeshkumar		

Local Organization

Andrea Bandini and Elena Della Godenza at Centro Congressi di Bertinoro.

Sponsoring Institutions

Bertinoro International Center for Informatics (BICI)

Table of Contents

Conservation of Combinatorial Structures
in Evolution Scenarios

Sèverine Bérard[1], Anne Bergeron[2], and Cedric Chauve[2]

[1] LIRMM, Montpellier, France
`berard@lirmm.fr`
[2] LaCIM, Université du Québec à Montréal, Canada
{`anne,chauve`}`@lacim.uqam.ca`

Abstract. This paper investigates the problem of conservation of combinatorial structures in genome rearrangement scenarios. We give a characterization of a class of scenarios that conserve all common intervals, called commuting scenarios, and a characterization of permutations for which commuting scenarios exist. We show that measuring conservation of common intervals can be useful tool in assessing the quality of rearrangement scenarios, by investigating in detail three specific scenarios involving the mouse, rat and human X chromosomes.

1 Introduction

The reconstruction of evolution scenarios based on genome rearrangements has proven to be a powerful tool in understanding the evolution of close species, especially mammals. For example, in the last two years, several very interesting evolution scenarios have been proposed between the mouse and the human [15], and the between human, the mouse and the newly sequenced Norway rat [7, 10], using the MGR and GRIMM software [6, 17].

MGR relies heavily on *sorting signed permutations by inversions* and the related *median* problem [14]. However, the number of parsimonious sequences of inversions can be exponential [3]. It is then natural to ask for some additional criteria that can help to select putative scenarios. We are interested in scenarios that do not break combinatorial structures, defined in terms of genomic segments, that are conserved in both chromosomes. Indeed, if two genomes share a common feature, it is likely that their common ancestor did too, which makes evolution scenarios that conserve this feature interesting.

In this work, the combinatorial structures that we consider are *common intervals* [18, 12]. We give a characterization of a class of evolution scenarios between two chromosomes that are both parsimonious and do not break any interval of genomic segments that is common to the two chromosomes. We call this class of scenarios *commuting scenarios*, and we describe a linear time algorithm to decide if a commuting scenario exists, and compute it if it does exist.

Sections 2 and 3 present the basic concepts and definitions. In Section 4, we discuss results on human, mouse and rat chromosomes X [10], highlighting the role of common intervals in assessing the quality of evolution scenarios. The main theoretical results are presented in Section 5.

J. Lagergren (Ed.): RECOMB 2004 Ws on Comparative Genomics, LNBI 3388, pp. 1–14, 2005.
© Springer-Verlag Berlin Heidelberg 2005

2   Rearrangement Scenarios

A *signed permutation* is a permutation on the set of integers $\{0, 1, 2, \ldots, n\}$ in which each element has a sign, positive or negative. An *inversion* of an interval of a signed permutation inverts the order of the elements of the interval, while changing their signs. In the following, we will assume that genomes are modeled by signed permutations, and that rearrangement operations are restricted to inversions. A *rearrangement scenario* between two or more genomes is given by an unrooted tree whose nodes are labeled by permutations and such that each of the given genomes labels a leaf, and the permutations labeling two adjacent nodes differ by one inversion.

The number of vertices of the tree is one more the number of rearrangements of the scenario. A scenario with a minimum number of rearrangements is called an *optimal* scenario. For example, given the three permutations $G_1 = (1\ 2\ 3\ 4\ 5\ 6)$, $G_2 = (1\ \bar{3}\ \bar{2}\ 5\ 4\ 6)$, $G_3 = (1\ \bar{5}\ \bar{2}\ \bar{4}\ 3\ 6)$, two rearrangement scenarios for G_1, G_2 and G_3 are proposed in Fig. 1, each of them having 6 rearrangements.

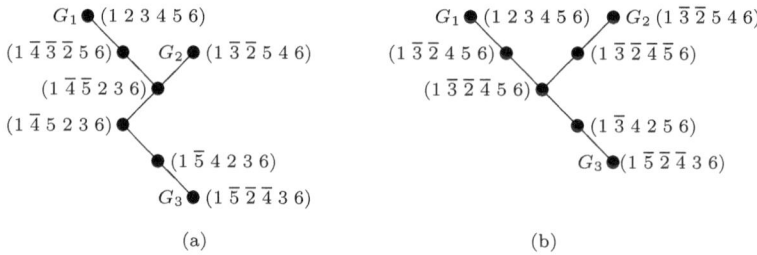

(a)   (b)

Fig. 1. Two rearrangement scenarios between permutations G_1, G_2 and G_3.

With two permutations, there exist polynomial algorithms to compute an optimal scenario [11, 13, 2, 16], but the problem becomes NP-hard for more than two permutations [8], although good heuristics are available [14].

Usually, there is more than one optimal scenario, even with different tree topologies. For example, Fig 1(b) gives an alternate scenario for genomes G_1, G_2 and G_3, that yields a different *median*, or *common ancestor*, for the three species. Is one scenario "better" than the other? When dealing with real genomes, only further insight from biology and evolution history will allow to completely settle this question. However, we will show that tracking some combinatorial structures along different scenarios can help to partially assess the quality of a scenario. For example, using data from chromosomes X of the mouse, human and rat [10], we were able to detect a major difference between two mouse assemblies – a transposition of two blocks of more than 300k base pairs.

3   Common Intervals and Commuting Inversions

A *point* $p \cdot q$ in a permutation is defined by a pair of consecutive elements in the permutation. When a point is of the form $i \cdot i + 1$, or $-(i + 1) \cdot -i$, it is called an *adjacency*, otherwise it is called a *breakpoint*.

An interval of a permutation is defined either by giving its *endpoints*, or by giving the set of its (unsigned) elements $\{|p_i|, \ldots, |p_j|\}$. A non-empty interval of the identity permutation can also be specified by giving its first and last element, such as $[i..j]$, in which case it is then understood that all elements between i and j belong to the interval.

The notion of *common interval* was studied among others in [12] in order to model the fact that a group of genes can be rearranged in a genome but still remain connected.

Definition 1. *A* common interval *of two signed permutations P and Q is a set of two or more integers that is an interval in both permutations.*

For example, the common intervals of permutations $G_2 = (1\ \overline{3}\ \overline{2}\ 5\ 4\ 6)$ and $G_3 = (1\ \overline{5}\ \overline{2}\ \overline{4}\ 3\ 6)$ are $\{2,5\}$, $\{2,4,5\}$, $\{2,3,4,5\}$, $\{1,2,3,4,5\}$, $\{2,3,4,5,6\}$, and $\{1,2,3,4,5,6\}$.

Definition 2. *Two distinct sets of integers A and B are said to* commute *if they trivially intersect, that is, $A \subset B$, or $B \subset A$, or $A \cap B = \emptyset$.*

The above definition, together with the fact that both intervals and inversions can be represented as sets of integers, is central in the links we establish between inversions, commutation and conservation of intervals. Indeed, an inversion that commutes with an interval does not change the composition of this interval, while it may change the order of the elements within the interval.

Definition 3. *The* score *of a scenario between two signed permutations P and Q is the ratio of inversions that commute with all common intervals of P and Q over the number of inversions in the scenario.*

For example, each of the rearrangement scenarios of Fig. 1 induces three scenarios with two permutations. The scores are given in Table 1, and show that the second scenario is slightly better, in terms of conservation, than the first.

Table 1. Conservation scores of the two evolution scenarios of Fig. 1.

	Scenario (a)	Scenario (b)
G_1 to G_2	0/3	4/4
G_1 to G_3	2/5	2/4
G_2 to G_3	2/4	2/4
Total	4/12	8/12

Efficient computation of scores is based on the notion of *irreducible intervals*. A common interval I between P and Q is an irreducible interval if there is an adjacency of P contained in I, and I is the smallest common interval between P and Q that contains this adjacency. For example, irreducible intervals between G_1 and G_2 of Fig. 1 are: $\{1,2,3\}$ (for adjacency $1 \cdot 2$), $\{2,3\}$ (for adjacency $2 \cdot 3$), $\{2,3,4,5\}$ (for adjacency $3 \cdot 4$), $\{4,5\}$ (for adjacency $4 \cdot 5$), and $\{4,5,6\}$ (for adjacency $5 \cdot 6$). Any common interval is the union of a sequence of irreducible intervals such that two consecutive intervals intersect. This implies:

Proposition 1. *An inversion breaks a common interval if and only if it breaks an irreducible interval.*

Since, there are at most n irreducible intervals between two signed permutations of $\{1, 2, \ldots, n\}$, and they can be identified in linear time [12], Proposition 1 implies that computing scores can be done efficiently.

4   Reconstructing the Ancestral Chromosome X

The scenario proposed in [10], that attempts to reconstruct a putative chromosome X for the common ancestor of man, mouse and rat, motivated our investigations in conservation of common intervals. The data, based on the conservation of *synteny blocks*, yields permutations on 16 integers for each species.

This scenario, displayed in Fig. 2, has a remarkable feature: there is a common interval between the three species, $\{5, 6\}$, that is not conserved in the intermediate permutations!

```
Human  1  2  3  4 [5][6] 7   8   9  10  11  12 13 14 15 16
       1  2  3  4 [5][6] 7   8   9  10  11  12 13 14 15̄ 16
       1  3̄ 2  4 [5][6] 7   8   9  10  11  12 13 14 15̄ 16
       1  3̄ 2 [5̄] 4̄ [6] 7   8   9  10  11  12 13 14 15̄ 16
       1  3̄ 2 [5̄]12̄ 11̄ 10̄  9̄  8̄  7̄ [6̄] 4 13 14 15̄ 16
Median 1  3̄ 9 10 11 12 [5]  2   8̄  7̄ [6̄] 4 13 14 15̄ 16

Mouse [5̄][6̄] 4 13 14 15̄ 16 1  3̄  9 10̄ 11 12  7̄ 8 2̄
      [5̄][6̄] 4 13 14 15̄ 16 1  3̄  9 10  11 12  7̄ 8 2̄
      [5̄][6̄] 4 13 14 15̄ 16 1  3̄  9 10  11 12  7  8 2̄
      [5̄]12̄ 11̄ 10̄ 9̄ 3 1̄ 16̄ 15 14̄ 13̄ 4̄ [6] 7 8 2̄
       1  3̄ 9 10 11 12 [5]16̄ 15 14̄ 13̄ 4̄ [6] 7 8 2̄
Median 1  3̄ 9 10 11 12 [5] 2   8̄  7̄ [6̄] 4 13 14 15̄ 16

Rat   13̄ 4̄ [5][6̄]12̄ 8̄ 7̄ 2 1 3̄ 9 10 11 14 15̄ 16
      13̄ 4̄ [6][5̄]12̄ 8̄ 7̄ 2 1 3̄ 9 10 11 14 15̄ 16
      13̄ 4̄ [6] 7  8 12 [5] 2 1 3̄ 9 10 11 14 15̄ 16
      13̄ 4̄ [6] 7  8  2̄ [5̄]12̄ 1 3̄ 9 10 11 14 15̄ 16
      13̄ 4̄ [6] 7  8  2̄ [5̄]12̄ 11̄ 10̄ 9̄ 3 1̄ 14 15̄ 16
Median 1  3̄ 9 10 11 12 [5] 2  8̄  7̄ [6̄] 4 13 14 15̄ 16
```

Fig. 2. The evolution scenario for human, mouse and rat X Chromosome of [10].

The first column of Table 2 gives the scores for the three corresponding pairwise scenarios, which are very low, even as low as 2/10 in the scenario transforming the rat chromosome X into the human chromosome X. The loss of the common interval $\{5, 6\}$ in the intermediate species clearly has a damaging effect on the scores. It also induces three independent reconstructions of this interval along the three branches that go from the median ancestor to the human, to the mouse, and to the rat.

In such a situation, it is possible to question several hypothesis of the model: the parsimony assumption, the evolutionary model, that considers only inversions, or, more simply, the data. Questioning the data was the easiest experiment, since the positions of the blocks were available. We soon realized that the mouse assembly used to construct the data, (assembly 30 of UCSD), differed from a more recent version, (assembly 32 of UCSD) notably on the respective position of synteny blocks 5 and 6, that are transposed[1].

Using the same synteny blocks in the order given by assembly 32 of mouse, and MGR we obtained an alternate scenario, with much better scores, as the second column of Table 2 shows: the total score increased from 10/30 to 18/30. Finally, we defined our own data set, that resulted in three permutations on 22 elements, as follows: we considered the genes of chromosome X in human (assembly 35.1 of NCBI), mouse (assembly 33.1 of NCBI) and rat (assembly 2.1 of NCBI), and we identified the genes common to the three genomes on the basis of their functional annotation, by using both confirmed and predicted annotations. The scenario we computed on this data set, described in Appendix A, is displayed in Figure 3, and the corresponding scores in the third column of Table 2.

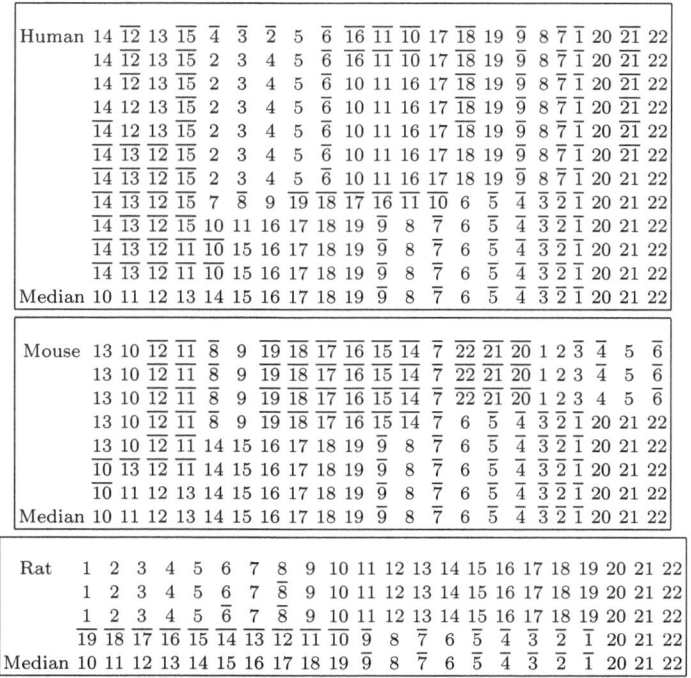

Fig. 3. The scenario from our 22 blocks data set.

The conservation score we proposed and illustrated in this section is a first attempt to measure the conservation of combinatorial structure in evolution

---

[1] The mouse assembly 33 of UCSD agrees with the positions given in assembly 32.

Table 2. Conservation scores of the scenario presented in [10], of a new scenario using the assembly 32 of the mouse, and of the scenario based on our gene blocks.

	Mouse 30	Mouse 32	Mouse 32 + 22 blocks
Human to Mouse	4/10	8/10	15/18
Human to Rat	2/10	6/10	13/15
Rat to Mouse	4/10	4/10	5/11
Total	10/30	18/30	33/44

scenarios, but raises many interesting questions. For example, the possibility to weight inversions that break common interval with respect to the position of the corresponding edge in the tree, or with respect to the number of broken intervals, should be considered. The interpretation of scores on permutations of different sizes, as we have in our example, is another interesting question.

5   Perfect Scenarios

A *perfect scenario* is a scenario in which no inversion breaks a common interval. The construction of perfect scenarios is discussed in [9], where the problem is shown to be computationally difficult. Perfect scenarios always exist between two permutations, but are not necessarily optimal. They can even be trivial when the two permutations have few common intervals. For example, any scenario between permutations (1 2 3 4) and (3 1 4 2) is perfect. However, permutations that arise from genomic data of relatively close species share lots of common intervals, and some rearrangement scenarios have striking features in terms of structure conservation.

In this section, we study a class of perfect scenarios, called *commuting scenarios*, and we show that deciding the existence of optimal commuting scenario, and constructing them, can be done in linear time.

5.1   Commuting Scenarios

Definition 4. *Let $r_1, \ldots, r_k$ be a sequence of inversions that transforms a permutation P into a permutation Q. The sequence $r_1, \ldots, r_k$ is a commuting scenario if, for every $i, j \in [1..k]$, the inversions r_i and r_j commute and are distinct.*

A beautiful example of a commuting scenario, is given in [7] where a region on human chromosome 17 (denoted by H below) is compared to a region on mouse chromosome 11 (denoted by M). The resulting permutations are:

$$H = (1\ 2\ 3\ 4\ 5\ 6\ 7\ 8\ 9\ 10\ 11\ 12\ 13\ 14\ 15\ 16\ 17\ 18\ 19),$$
$$M = (7\ \overline{8}\ 6\ \overline{5}\ 4\ \overline{3}\ 1\ 2\ \overline{10}\ 9\ 11\ 12\ \overline{13}\ 16\ \overline{15}\ 14\ 17\ \overline{18}\ 19),$$

The mouse chromosome M can be obtained from chromosome H by the commuting scenario of Fig. 4, in which all the inversions are identified by underlining the set of inverted integers. Note also that we choose to represent the scenario

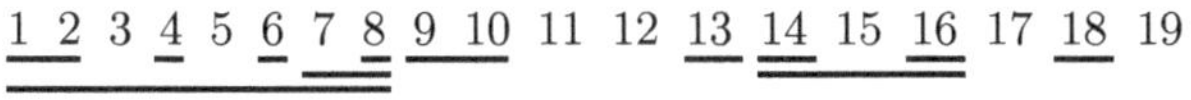

1 2 3 4 5 6 7 8 9 10 11 12 13 14 15 16 17 18 19

Fig. 4. A commuting scenario transforming H into M.

by inversions applied to the identity permutation. This will be helpful in proving and understanding properties of commuting scenarios.

The fact that one could simultaneously underline all inversions in Fig. 4 is a direct consequence of the fact that all inversions commute, and implies the following lemma. This example highlights many properties of commuting scenarios. For example, applying the inversion from the largest to the smallest transforms H into M by always inverting segments of H composed of consecutive integers. More important for us is the following lemma.

Lemma 1. *Let S be an optimal commuting scenario between two permutations P and Q. An interval I is a common interval of P and Q if and only if all inversions of S commute with I.*

Proof. First, it is immediate that if each inversion of S commute with I, then I is a common interval between P and Q.

On the other hand, since S is a commuting scenario, one can first apply all inversions that commute with I, which leads to a permutation P' in which I is an interval. Let R be the set of remaining inversions, those that do not commute with I. If all inversions in R are disjoint, there are at most two of them, and it is easy to see that applying them to P' yields a permutation in which I is no longer an interval.

Suppose R contains at least two non-disjoint intervals, and that these intervals intersect I at its right extremity – the argument is completely symmetrical if we consider the left extremity. Let r be the largest interval that intersect I at its right extremity, s the second largest, and I' the non-empty set of elements of I that are at the left of r. Since the scenario is optimal, s is strictly contained in r. Therefore, r can be partitioned into disjoint intervals, u, s and v, such that u is contained in I, v is disjoint from I, and at least one of u and v is non-empty.

Applying both r and s to P' exchange the intervals u and v, leaving s in the middle. By the choice of r and s, this structure will remain unchanged for the rest of the sorting procedure, except for possible inversions within s. Note also that all elements of I' will remain to the left of r. If u is not empty, the elements of s that are not in I will end up between I' and u. If v is not empty, the elements of v will end up between I' and $s \cap I$. Thus I is not an interval in Q, and cannot be a common interval of P and Q. □

Proposition 2. *Optimal commuting scenarios between two permutations P and Q conserve all common intervals of P and Q in all intermediate permutations.*

Proof. This follows immediately from Lemma 1. □

The scenario of Fig. 4 is also an optimal scenario. It is not true, in general, that if an optimal commuting scenario exists, then all optimal scenarios are commuting. An example is again given by data from human and mouse. Consider the first 8 segments of H and M: one can transform H into M with the sequence of inversions of Fig. 5, that is an optimal scenario, but not a commuting one.

$$
\begin{array}{ccccccccc}
1 & 2 & 3 & \underline{4} & 5 & \underline{6} & 7 & 8 \\
\hline
1 & 2 & \overline{7} & 3 & \overline{4} & 5 & \overline{6} & 8 \\
\hline
7 & \overline{2} & \overline{1} & 3 & \overline{4} & 5 & \overline{6} & 8 \\
7 & \overline{8} & 6 & \overline{5} & 4 & \overline{3} & 1 & 2
\end{array}
$$

Fig. 5. A non commuting scenario transforming the first 8 segments of H into M.

Deciding whether an optimal commuting scenario exists is therefore not a trivial question. We give, in the following section, a characterization of permutations that admit optimal commuting scenarios.

5.2   Existence of Optimal Commuting Scenarios

The results of this section rely on many of the concepts that have been developed around the sorting by inversion problem. The terminology and conventions we follow are presented in [5].

Note. In this section we consider signed permutations of $\{0, 1, \ldots, n\}$ that start with 0 and end with n.

A first remark is that, since an optimal commuting scenario does not break any common interval by Proposition 2, such a scenario can only exist for permutations that can be optimally sorted component by component. We first settle the case of oriented components in Theorem 1 that relates three fundamental types of intervals: common intervals, inversions of a commuting scenario, and the elementary intervals of the sorting by inversion theory, that appear here as the vertices of the overlap graph.

Theorem 1. *Let C be an oriented component of a permutation. The three following statements are equivalent.*

1. *C can be sorted by an optimal commuting scenario.*
2. *The overlap graph of C is a tree.*
3. *C can be sorted by an optimal scenario in which each inversion is a common interval.*

Before proving Theorem 1, we establish some properties of overlap graphs.

Lemma 2. *Any leaf of an overlap graph is an oriented interval, and a common interval.*

Proof. A leaf is an interval that overlaps only one other interval. Let m and M be the minimal and maximal values of a non-empty interval I_p. Then I_p

always overlaps both I_{m-1} and I_M. If I_p is a leaf, we must have either $p = M$ or $p = m - 1$, implying in both cases that the interval is oriented, since it contains exactly one of its extremities. If I_p is not a common interval, then $|I_p| \geq 2$ and I_p does not contain all the integers between m and M. Let m' and M' be, respectively, the smallest and largest missing integers. Note that one can have $m' = M'$. Then I_p overlaps $I_{m'-1}$ and $I_{M'}$, and therefore can not be a leaf.   □

Lemma 3. *Erasing a leaf of an overlap graph that is a tree is always a sorting inversion.*

Proof. Erasing a leaf never disconnects a tree, and we only have to check that the tree contains at least another oriented interval after the leaf has been erased. If a leaf I_q overlaps I_p, and I_p is unoriented, then applying I_q orients I_p. If I_p is oriented and is the only other leaf (i.e, the graph contains only two vertices I_p and I_q), then $I_p = I_q$, and the component is sorted after erasing I_q. If I_p is connected to exactly one other node, I_p would become an unoriented leaf in the new tree, which is impossible by Lemma 2. Thus I_p is connected to at least two other nodes, implying at least two other leaves, therefore two other oriented intervals.   □

Proof (Theorem 1).

   (1) $\implies$ (2). Suppose that an oriented component can be sorted by an optimal commuting scenario. We will show, by induction on the length of the scenario, that it must be a tree. It is certainly true for scenarios of length 1, since the overlap graph has two nodes. We will show that there always exists an inversion r that does not contains any other inversion of the scenario, and that necessarily creates an adjacency. Since r can be applied first, this implies that r is an elementary interval and a leaf in the overlap graph, and the result follows by induction. If all inversions are disjoint, then the leftmost certainly creates an adjacency. Otherwise, let r be a smallest inversion, in terms of the number of elements it inverts, included in another inversion, and s the smallest inversion containing r. As the inclusion of r in s is strict, then r creates an adjacency between one of its elements and an element of s not included in r.

   (2) $\implies$ (3). It follows immediately from Lemmas 2 and 3: erasing a leave always yields a tree.

   (3) $\implies$ (1). Let S be a set of reversals that transforms permutation P into perrmutation Q, and such that all inversions of S are common intervals of P and Q. Apply to P a maximal subset R of commuting inversions from S yielding permutation P'. By Lemma 1, all common intervals of P and P' commute with all inversions of R. Let R' be the set of remaining inversions. If R' is not empty, there is at least one inversion s that is an interval of P', and that does not commute with an inversion of R. Therefore, s cannot be an interval of P, and is not a common interval of P and Q.   □

   In order to have a characterization of all permutations that admits optimal commuting scenarios, we must next deal with unoriented components. In this case, one inversion is allowed to orient the component, but, as we saw in Theorem 1, the overlap graph of the resulting component must be a tree, which restricts severely the structure of the overlap graph of the original unoriented component: it must contain only one cycle. This imposes to unoriented components the following *reduced* form[2].

Theorem 2. *An unoriented component admits an optimal commuting scenario if and only if it can be reduced to a permutation of the form*

$$(0 \ \ 2k \ \ 2k-1 \ \ ... \ \ 3 \ \ 2 \ \ 1 \ \ 2k+1).$$

Proof. Let P be a positive permutation P with n breakpoints and one component. If P can be optimally sorted with a commuting scenario $\mathcal{S}$, then each inversion of $\mathcal{S}$ can be applied first, and must create a permutation whose overlap graph is a tree. We will show that the inversions of $\mathcal{S}$ are either single elements, or all the elements of the interval $[1..n-1]$, implying that P is of the form $(0 \ \ n-1 \ \ n-2 \ \ ...3 \ \ 2 \ \ 1 \ \ n)$, and that the length of a commuting scenario is n.

Let r be an inversion of $\mathcal{S}$, with minimum element m and maximum element M, then applying r to P creates the oriented elementary intervals I'_{m-1} and I'_{M}, that are leaves of the resulting overlap graph. By Lemmas 1 and 2, those intervals must commute with r, implying that r also commutes with I_{m-1} and I_M.

If r commutes with I_{m-1}, then either $m-1$ is immediately to the left of interval r, or m is the first element of r. Similarly, if r commutes with I_M, then either $M+1$ is immediately to the right of interval r, of M is the last element of r.

If m is the first element of r, and M the last, interval r will be a component unless $m = M$. This case produces the inversions consisting of a single element. If $m-1$ is immediately to the left of r, and $M+1$ is immediately to the right, then $[m-1..M+1]$ is a component, thus $m-1 = 0$, and $M+1 = n$. This case produces the inversion of the interval $[1..n-1]$.

Finally, if $m-1$ is immediately to the left of r, and M the last element of r, then $[m-1..M]$ is a component, implying that $M = n$, which is impossible. Similarly, $M+1$ is immediately to the right of r, and m the first element of r, implies $m = 0$, which is also impossible.

Applying all the possible n commuting inversions to the identity permutation yields the permutation: $(0 \ \ n-1 \ \ n-2 \ \ ...3 \ \ 2 \ \ 1 \ \ n)$. If n is odd, the inversion distance is n, thus the permutation can be sorted by an optimal commuting scenario. However, if n is even, the permutation has two cycles, and there exists an optimal scenario of length $n-1$. □

5.3   Algorithm

We now have all the elements needed to construct a linear time algorithm that will decide if a permutation P of size n can be sorted with an optimal commuting scenario, and that will compute the necessary information to obtain such a scenario, if it exists.

---

[2] A component is *reduced* if all the smaller component contained in it have been sorted, and all the resulting adjacencies collapsed into single elements [3].

General overview. The overlap graph can be processed component by component, and Theorem 2 addresses the case of unoriented components.

In the case of an oriented component, whose overlap graph C has k vertices, the algorithm given in Fig. 6 decides, in linear time, if C can be sorted by a commuting scenario. If such a scenario exists, the algorithm computes, in linear time, the order in which the vertices of the overlap graph must be erased to produce a commuting set of inversions.

---

1. Build iteratively the edges of C as long as there are at most $k - 1$.
2. If C has at least k edges, then C is not a tree.
3. Else, remove iteratively the leaves of C, which produces the sequence of inversions necessary to sort C.

Fig. 6. Algorithm 1 (Main algorithm).

As one can see, the core of this algorithm is step 1., that ensures that one can decide if the component can be sorted with a commuting scenario after considering at most k edges. The last step, that produces the scenario, can clearly be done in a single traversal of the overlap tree, and thus in $O(k)$ time. So we need to describe how we identify at most k edges of the overlap graph in time $O(k)$.

Computing edges of the overlap graph. Let (ℓ_i, r_i) be the indices, respectively, of the left point and right point of the elementary interval I_i of C. The first step is to compute a sequence S of the $2k$ ℓ_i's and r_i's in such a way that the following property holds: two intervals I_p and I_q overlap if and only if in S ℓ_q appears between ℓ_p and r_p and r_q appears after r_p. This can be done in $\Theta(k)$ worst-case time.

Once the sequence S is built, the following algorithm computes at most m edges of the overlap graph during a single pass on S. One denotes by S_i the ith element of S and m the maximum number of edges one wants to produce.

---

1. Let $i = 1$.
   // *Invariant: there are only ℓ_q's at the left of S_i, and all of them have their corresponding r_q's at the right of S_i or at S_i.*
2. While $i \leq 2n$ and less than m edges have been computed do
   3. If $S_i = r_p$ for some p then
      4. Let $S_j = \ell_p$.
      5. For every ℓ_q located between S_j and S_i do add an edge (p, q).
      6. Remove from S the elements ℓ_p and r_p.
         // *This last step ensures the invariant still holds*

Fig. 7. Algorithm 2 (Computing edges of the overlap graph).

Using the appropriate data structures to encode S, such as a double-linked list, one can implement this algorithm in order that instruction 6, that removes elements of S, is done in $\Theta(1)$ time, and that instruction 5, that visits all elements between ℓ_p and r_p, is done in time proportional to the number of these elements. The invariant ensures that the elements visited between ℓ_p and r_p are exactly the ℓ_q's such that the corresponding r_q's are located after r_p. This leads to the following lemma:

Lemma 4. *For every m, Algorithm 2 computes at most m edges of C in $\Theta(m + k)$ worst-case time.*

Theorem 3. *It can be decided, in $\Theta(n)$ time and space, whether a signed permutation P on n elements can be sorted by an optimal commuting scenario. If an optimal commuting scenario exists, one can compute the corresponding sequence of oriented inversions in $\Theta(n)$ time and space.*

Proof. Given P, we can process it component by component. The case of unoriented components, that can be detected in $\Theta(n)$ time [4], has been addressed in Theorem 2 and can be solved in $\Theta(n)$ time. Next, one needs to compute the set of vertices of each component of the overlap graph, and this can be done in linear time using [4, 1]. Then we apply Algorithm 1, where step 1 is done with Algorithm 2, and step 3, if necessary, is done during a depth-first traversal of the overlap graph of C where each leaf is processed – the corresponding inversion is added to the scenario – during its first visit. Steps 1 and 3 take $\Theta(k)$ time if the current component has k vertices, by definition of a depth-first traversal and Lemma 4, which leads to a total $\Theta(n)$ time complexity to build the forest of trees that composes the overlap graph and iteratively remove the leaves of this forest.   □

6   Conclusion

We described in this paper a class of perfect scenarios, the commuting scenarios, and we showed that one can decide in linear time whether a signed permutation can be sorted by an optimal commuting scenario. However since a perfect scenario is not necessarily commuting, it is still an open question to decide in polynomial time if a permutation can be sorted by an optimal perfect scenario. It would also be interesting to have more information on how large is the class of permutations that can be sorted by commuting scenarios. This would help in assessing the significance of optimal commuting scenarios with respect to other optimal scenarios. Here, we focused on optimal commuting scenarios, and the class of non-optimal commuting scenarios should be investigated. Indeed, every permutation can be sorted by a perfect scenario, but it is not true that every permutation can be sorted by a commuting scenario. Finally, it should also be noted that the best time complexity for computing an optimal scenario for a general permutation is currently $(n\sqrt{n \log(n)})$ [16]. Our algorithm is the first, as far as we know, that achieves linear-time complexity for a non-trivial class of signed permutations.

References

1. D. A. Bader, B. M. E. Moret, and M. Yan. A linear-time algorithm for computing inversion distance between signed permutations with an experimental study. *J. Comp. Biol.*, 8(5):483–491, 2001.
2. A. Bergeron. A very elementary presentation of the hannenhalli-pevzner theory. In *CPM 2001*, volume 2089 of *Lecture Notes in Comput. Sci.*, pages 106–117, 2001. (Extended version to appear in Discrete Applied Math.).
3. A. Bergeron, C Chauve, T. Hartman, and K. St-Onge. On the properties of sequences of reversals that sort a signed permutation. In *JOBIM 2002*, pages 99–108, 2002.
4. A. Bergeron, S. Heber, and J. Stoye. Common intervals and sorting by reversals: A marriage of necessity. *Bioinformatics*, 18(Suppl. 2):S54–S63, 2002. (ECCB 2002).
5. A. Bergeron, J. Mixtacki, and J. Stoye. Reversal distance without hurdles and fortresses. In *CPM 2004*, volume 3109 of *Lecture Notes in Comput. Sci.*, pages 388–399, 2004.
6. G. Bourque and P. A. Pevzner. Genome-scale evolution: Reconstructing gene orders in the ancestral species. *Genome Res.*, 12(1):26–36, 2002.
7. G. Bourque, P. A. Pevzner, and G. Tesler. Reconstructing the genomic architecture of ancestral mammals: Lessons from human, mouse, and rat genomes. *Genome Res.*, 14(4):507–516, 2004.
8. A. Caprara. Formulations and hardness of multiple sorting by reversals. In *RE-COMB 1999*, pages 84–94, 1999.
9. M. Figeac and J.-S. Varré. Sorting by reversals with common intervals. In *WABI 2004*, volume 3240 of *Lecture Notes in Bioinformatics*, pages 26–37, 2004.
10. R. A. Gibbs et al. Genome sequence of the brown norway rat yields insights into mammalian evolution. *Nature*, 428:493–521, 2004.
11. S. Hannenhalli and P. A. Pevzner. Transforming cabbage into turnip: Polynomial algorithm for sorting signed permutations by reversals. *J. ACM*, 46(1):1–27, 1999.
12. S. Heber and J. Stoye. Finding all common intervals of k permutations. In *CPM 2001*, volume 2089 of *Lecture Notes in Comput. Sci.*, pages 207–218, 2001.
13. H. Kaplan, R. Shamir, and R. E. Tarjan. A faster and simpler algorithm for sorting signed permutations by reversals. *SIAM J. Computing*, 29(3):880–892, 1999.
14. B. M. E. Moret, A. C. Siepel, J. Tang, and T. Liu. Inversion medians outperform breakpoint medians in phylogeny reconstruction from gene-order data. In *WABI 2002*, volume 2452 of *Lecture Notes in Comput. Sci.*, pages 521–536, 2002.
15. P. A. Pevzner and G. Tesler. Genome rearrangements in mammalian evolution: Lessons from human and mouse genomes. *Genome Res.*, 13(1):37–45, 2003.
16. E. Tannier and M.-F. Sagot. Sorting by reversals in subquadratic time. In *CPM 2004*, volume 3109 of *Lecture Notes in Comput. Sci.*, pages 1–13, 2004.
17. G. Tesler. GRIMM: genome rearrangements web server. *Bioinformatics*, 18(3):492–493, 2002.
18. T. Uno and M. Yagiura. Fast algorithms to enumerate all common intervals of two permutations. *Algorithmica*, 26(2):290–309, 2000.

Appendix A

The 22 gene blocks of the new human/mouse/rat scenario

We give here the list of the first gene of the 22 blocks of genes we used in our study of the human/mouse/rat X chromosome evolution. The below table has the following format: the first column gives the integers associated to the 22 genes we consider, the next three columns present the rat data (gene/locus name, orientation and position) followed by three columns for the mouse and three columns for the human (assembly 34.3 for the human genome, assembly 32.1 for the mouse genome and assembly 2.1 for the rat genome).

1	Birc4	-	3013140	Birc4	+	33413632	BIRC4	+	121691803
2	Syn1	+	12517799	Syn1	-	19413738	SYN1	-	46478245
3	Sytl5	-	24886677	Sytl5	-	8475502	SYTL5	+	36924044
4	Cybb	-	25568041	Cybb	-	7985479	CYBB	+	36670221
5	Ebp	+	26384668	Ebp	-	6745931	EBP	+	47426239
6	Gdf-9b	-	29057402	Bmp15	-	4973481	BMP15	+	49570590
7	Pls3	-	29923507	Pls3	-	65568249	PLS3	+	113559762
8	Il13ra2	-	30559317	Il13ra2	-	135823846	IL13RA2	-	113002791
9	Dcx	+	34649731	Dcx	-	132129216	DCX	-	109300978
10	Ragb	-	38406475	LOC245670	+	141432864	RRAGB	+	54711109
11	Pfkfb1	+	39892044	Pfkfb1	+	138292880	PFKFB1	-	53926381
12	LOC317435	+	42360000	APXL	+	140920000	APXL	-	9250000
13	Mid1	-	44805419	Mid1	+	157816113	MID1	-	9827653
14	LOC302711	+	63450000	Tbl1x	+	67700000	Tbl1x	+	8845000
15	Dmd	+	71574635	Dmd	+	73462214	DMD	-	30498771
16	Maged1	-	82127336	Maged1	-	84274362	MAGED1	+	50553552
17	Arhgef9	-	82667648	Arhgef9	-	84771512	ARHGEF9	-	61721639
18	Slc16a2	-	91773989	Slc16a2	-	93602023	SLC16A2	+	72507876
19	Atrx	-	93979545	Atrx	-	95705534	ATRX	-	75517065
20	Xpnpep2	+	134548393	Xpnpep2	+	39428571	XPNPEP2	+	127578549
21	LOC367956	+	147367900	Gm366	+	51313500	LDOC1	-	138963500
22	Fmr1	+	154829464	Fmr1	+	58282553	FMR1	+	145661196

This data set induces the three following permutations (repectively, from top to bottom, for the rat, the mouse and the human), where the mouse chromosome X is represented reversed, in order to correspond to the evolution scenario given in Appendix C:

$$\text{Rat} = (\ 1\ \ 2\ \ 3\ \ 4\ \ 5\ \ 6\ \ 7\ \ 8\ \ 9\ \ 10\ 11\ 12\ 13\ 14\ 15\ 16\ 16\ 18\ 19\ 20\ 21\ 22\ )$$

$$\text{Mouse} = (\ 13\ 10\ \overline{12}\ \overline{11}\ \overline{8}\ \ 9\ \ \overline{19}\ \overline{18}\ \overline{17}\ \overline{16}\ \overline{15}\ \overline{14}\ \ 7\ \ \overline{22}\ \overline{21}\ \overline{20}\ \ 1\ \ 2\ \ \overline{3}\ \ \overline{4}\ \ 5\ \ \overline{6}\ )$$

$$\text{Human} = (\ 14\ \overline{12}\ 13\ \overline{15}\ \ \overline{4}\ \ \overline{3}\ \ \overline{2}\ \ 5\ \ \overline{6}\ \ \overline{16}\ \overline{11}\ \overline{10}\ 17\ \overline{18}\ 19\ \ \overline{9}\ \ 8\ \ \overline{7}\ \ \overline{1}\ \ 20\ \overline{21}\ 22\ )$$

Toward a Phylogenetically Aware Algorithm for Fast DNA Similarity Search

Jeremy Buhler and Rachel Nordgren

Department of Computer Science and Engineering
Washington University, St. Louis, MO 63130, USA
{jbuhler,rkn2}@cse.wustl.edu

Abstract. High-throughput DNA sequencing is now producing collections of genomes from moderately or closely related organisms. Such a collection may be represented as a multiple alignment M of orthologous sequences, which induces a phylogenetic tree τ. Long-range genomic alignments with phylogenies have not yet found a prominent place in BLAST-like similarity search algorithms, though using them directly as databases can potentially yield more accurate and more informative alignments.

This work describes how to construct local alignments between a query and a multiple alignment in a way that explicitly uses a phylogenetic tree τ. We give an EM algorithm to find a locally optimal alignment when the location of the query on the tree τ is not known. An initial implementation of the method is tested on a large multiple alignment of sequences from eight vertebrate genomes.

1 Introduction

The advent of high-throughput DNA sequencing technology has enabled not only broad sequencing, covering the genomes of diverse model organisms, but also *deep sequencing* of multiple genomes within a clade of more closely related species. Deep sequencing is useful for recognizing common genomic features across a set of organisms, for identifying genetic innovations unique to particular subgroups of the set, and more generally for reconstructing the organisms' evolutionary history. Recent large-scale projects include sequencing of multiple budding yeasts [1], *Plasmodium* species [2], and vertebrates [3].

As deep sequencing becomes a more common mode of genomic investigation, there is mounting pressure to develop computational methods that better exploit collections of long homologous sequences. Much work to date [4–7] has focused on preparing long-range multiple alignments of orthologous sequences across species. Alignments produced by these methods are more than the sum of their parts; in particular, they can provide valuable evidence about the rate and topology of evolution in a group of species, in the form of inferred phylogenetic trees.

Long-range genomic multiple alignments offer opportunities to integrate phylogenetic information into existing biosequence analysis problems. For example, recent work in gene structure prediction [8, 9] has begun to exploit phylogenetic

J. Lagergren (Ed.): RECOMB 2004 Ws on Comparative Genomics, LNBI 3388, pp. 15–29, 2005.
© Springer-Verlag Berlin Heidelberg 2005

information explicitly to improve recognition of meaningful conservation. However, one opportunity that has not yet been systematically exploited is the use of multiple alignments with phylogenetic information as databases to improve similarity search tools such as BLAST [10, 11]. In particular, consider the following problem:

> Given a DNA query sequence q and a multiple alignment M of orthologous DNA sequences, with an induced phylogenetic tree τ, find all sufficiently high-scoring local alignments between q and M.

This problem is well-posed, with the exception of a specification for how to score local alignments between q and M. We expect that using M as the database for search should lead to more accurate alignments, provided the evolutionary model associated with M is accurate. Moreover, an alignment of q to M provides information on where to place q on the phylogeny for M, which is helpful for, e.g., recognizing the species that yielded q or for incrementally reconstructing a history of paralogization for a gene family.

Statistical approaches to alignment and scoring that do not explicitly consider the tree τ can be derived as special cases of the scores proposed by, e.g., Yona and Levitt [12] and Wang and Stormo [13]. Alternatively, the alignment M can be decomposed into its individual sequences, and multiple alignments can be reconstructed iteratively from q and these sequences (again without reference to τ) as in PSIBlast [11]. However, neither of these methods explicitly take advantage of the information summarized in τ.

In this work, we formulate an explicitly phylogenetic method to perform similarity search of a query sequence against a multiple sequence alignment. More specifically, we describe how to extend an ungapped seed alignment between query and database into a gapped local alignment, using the phylogeny as the basis for scoring. Because the placement of q on the tree τ is uncertain, we give an iterative algorithm to find a locally most likely local alignment of q to M. This compute-intensive algorithm can be accelerated through caching of intermediate results. We have built an initial software implementation of our method, which we call PhyLAT, the **Phy**logenetic **L**ocal **A**lignment **T**ool.

The remainder of this work is organized as follows. Section 2 defines the probability model for alignments, formally states the phylogenetic gapped extension problem, and briefly describes the generation of seed alignments for this problem. Section 3 presents an EM approach to derive a locally maximum-likelihood local alignment of q to M from a given starting point. Section 4 details the procedure used to place q on the tree τ and describes how to accelerate this step by caching intermediate results. Section 5 investigates the performance of our method as implemented in PhyLAT. Finally, Section 6 concludes and identifies key directions for improving the speed and accuracy of our implementation.

2   Definitions and Formal Problem Statement

2.1   Alignment Probability Model

A database M consists of a multiple alignment of n homologous DNA sequences. Not all sequences need be present at all positions of M – the alignment may contain small gaps caused by local mutation or large gaps caused by larger-scale block insertion, deletion, and rearrangement. The database M comes with an inferred phylogenetic tree τ, whose n leaves are the individual sequences of M. τ defines both a topology on these sequences and a set of branch lengths. We assume that τ is rooted and binary; hence, τ has $2n - 2$ branches e_i, each with a length ℓ_i. To convert these branch lengths to transition probabilities, we are also given a mutation model consisting of a rate matrix Q and a stationary distribution π for the residue frequencies in M.

A local alignment A between a DNA sequence q and the database M is a partial correspondence between the residues of q and the columns of M. A aligns some substring q' of q to some interval M' of M, possibly inserting gaps in M' or q'. The probability of the data given an alignment A is, as usual, computed under the assumption that aligned residues are homologous, while all others are unrelated. More specifically, there exists an *augmented tree* τ^*, which consists of the original τ plus one more branch leading to a leaf labeled with q, such that the residues aligned by each column of A arise by common descent along τ^*. The remaining positions of q are independent of M.

Assuming that the residues of q are stochastically independent, as are the columns of M, we derive

$$\Pr(q, M \mid A, \tau^*) = \prod_{j=1}^{|A|} \Pr(y[j], Z[j] \mid \tau^*) \cdot \prod_{y \notin q'} \Pr(y) \prod_{Z \notin M'} \Pr(Z \mid \tau) . \quad (1)$$

where y_j and Z_j (either of which may be a gap "$-$") denote the parts of the jth column of A drawn from q and M, respectively. As usual, the assumption of independence is not biologically most appropriate but is consistent with normal practice in pairwise alignment. The residues of q follow an i.i.d. model, while each column of M is determined independently by common descent from τ.

In this work, we treat gaps in either q or M as first-class residues when computing the probability of the data given A. The rate matrix Q is therefore 5×5, rather than the more common 4×4, to model insertion and deletion as well as substitution. PhyLAT's rate matrix Q is a time-reversible model similar to the extended Tamura-Nei [14] model described in [15]. It is built from a set of stationary frequencies and three free parameters, corresponding to instantaneous rates of transitions, transversions, and indels.

Treating gaps the same as other residues is extremely convenient for phylogeny but has some disadvantages, notably a tendency to assign too large a probability to alignments containing long gaps. To partially compensate for this tendency, we extend the probability model of Equation (1) to implement an affine penalty for new gaps introduced into M or q by our algorithm, as described in, e.g., [16, Chapter 2]. However, a more balanced treatment of gaps in PhyLAT remains a topic for future work.

2.2   Formulation of Alignment Problem

Suppose we have reason to believe that a high-scoring local alignment between q and M exists between some substring $q[a, b]$ of q and some interval $M[c, d]$ of M.

Fig. 1. The four possible trees created by augmenting a single tree τ on three leaves. Each tree adds a single branch (shown by a dashed line) off one of the four existing branches.

Typically, the window $q[a, b] \times M[c, d]$ is determined by finding a *seed alignment* using an efficient hashing strategy [11, 17]. Our goal is to investigate whether the proposed high-scoring alignment actually exists.

To score any given alignment of q to M, we need to know which augmented tree τ^* describes the relationship of q to the fixed tree τ on M. However, the correct choice of τ^* may be *unknown* and indeed may vary from one local alignment to the next. Allowing τ^* to be unknown and variable permits lack of global synteny between features in q and features in M and allows for the possibility that a search tool user may have limited knowledge of how the query relates to the contents of the database. To deal with ignorance of τ^* when scoring A, we can enumerate the $2n - 2$ potential augmented topologies $\tau_1^* \ldots \tau_{2n-2}^*$, as illustrated in Figure 1, and marginalize over the choice of unknown topology.

To summarize, PhyLAT's extension of seed alignments must solve the following problem:

Problem 1. Let M be a multiple alignment on n sequences over an alphabet Σ, with the sequences related by a phylogenetic tree τ with branch lengths, and let q be a query sequence, also over Σ. Let $\tau_1^* \ldots \tau_{2n-2}^*$ be the $2n - 2$ possible augmented tree topologies connecting q to τ.

Find a local alignment A between M and q, restricted to some window $q[a, b] \times M[c, d]$, that maximizes the probability

$$\Pr(q, M \mid \tau, A) = \sum_i \Pr(q, M, \tau_i^* \mid \tau, A) .$$

For compactness of notation, we generally drop the explicit dependence of $\Pr(q, M \mid \tau, A)$ on the fixed tree τ from now on.

2.3   Generation of Seed Alignments

Most existing approaches to seeded alignment work on pairs of sequences, rather than a sequence and a multiple alignment. To use these methods without change, PhyLAT applies them to q and α_M, the most likely ancestral sequence given M and τ. We use hashing-based seed generation followed by ungapped extension with a scoring matrix from the DNA-PAM-TT family [18]. Seed alignments that pass a given E-value threshold as determined by ungapped Karlin-Altschul statistics [19] become candidates for gapped extension with the full phylogenetic algorithm.

The local window $q[a, b] \times M[c, d]$ in Problem (1) is centered on the seed alignment. We restrict alignment to a band about the seed of fixed width w. The length of the band is determined dynamically: starting with a small fixed window size, we progressively double this size until doing so does not improve the alignment score. Alignments are constrained to pass through the column of M at the midpoint of the seed, so that expanding the window does not cause PhyLAT to ignore the seed in favor of some unrelated high-scoring alignment.

3   EM Computation of Most Likely Local Alignment

In this section, we derive an algorithm for Problem 1 using expectation maximization [20]. The algorithm iteratively refines an initial guess $A^{(0)}$ at the alignment A while simultaneously inferring a distribution over the position of q relative to the tree τ. EM ensures that the final alignment is locally maximal in the neighborhood of $A^{(0)}$.

We first define a set of $2n - 2$ indicator variables x_i for the augmented topology:

$$x_i = \begin{cases} 1 \text{ if augmented topology is } \tau_i^* \\ 0 \text{ otherwise.} \end{cases}$$

The mth iteration of the EM algorithm starts with a previous best local alignment $A^{(m-1)}$. In the E-step of the iteration, the algorithm computes for each x_i the expectation

$$\hat{x}_i = \Pr(x_i = 1 \mid q, M, A^{(m-1)}) . \tag{2}$$

In the M-step, the algorithm computes a new alignment $A^{(m)}$ to maximize the sum

$$\sum_i \hat{x}_i \log \Pr(q, M \mid x_i = 1, A^{(m)}) . \tag{3}$$

Each iteration of EM both increases the likelihood of $A^{(m)}$ given the data and updates the estimated probabilities of the trees τ_i^*, until a locally maximal alignment and a final distribution over augmented trees is reached.

Relative to the standard formulation of EM, the known data, missing data, and model parameters in this problem are respectively the sequence q and alignment M, the augmented tree selectors x_i, and the alignment A of q to M.

3.1   Computation of E-Step

To compute $\hat{x}_i$, we apply Bayes' theorem to Definition (2):

$$\hat{x}_i = \frac{\Pr(q, M \mid x_i = 1, A^{(m-1)}) \Pr(x_i = 1 \mid A^{(m-1)})}{\Pr(q, M \mid A^{(m-1)})} .$$

The first term of the numerator is given by Equation (1), while the second term is independent of the actual data and may be viewed as the *user's prior* over the position of q on the tree. This prior may be taken as uninformative (uniform) or may be weighted based on available information. For example, if the user knows that q comes from a mammal, then the position of q may be highly biased toward a particular subtree of a tree on vertebrates. The denominator is independent of i; rather than computing it explicitly, we observe that $\sum_i \hat{x}_i = 1$, since one of the augmented topologies must be correct, and normalize accordingly.

To further simplify the E-step, we observe that, according to Equation (1), the probability $\Pr(q, M \mid A^{(m-1)}, x_i = 1)$ can be decomposed as

$$C \cdot \prod_j \Pr(y_j, Z_j \mid \tau_i^*) ,$$

where j runs over the positions of $A^{(m-1)}$ and C is a constant independent of i, which can be normalized away along with the denominator. Hence, it is enough to compute

$$\hat{x}_i \propto \left(\prod_j \Pr(y_j, Z_j \mid \tau_i^*) \right) \Pr(x_i = 1 \mid A) , \tag{4}$$

which requires inspecting only the positions of the local alignment $A^{(m-1)}$. To ensure adequate numerical precision with long alignments A, the product in (4) should be computed in the log domain.

3.2   Computation of M-Step

We first observe that the alignment maximizing the sum (3) does not change if we subtract from this sum the constant

$$C' = \log \Pr(q) \Pr(M \mid \tau) = \sum_i \hat{x}_i \log \Pr(q) \Pr(M \mid \tau) .$$

Again applying the definition in Equation (1) and canceling common terms, we find that the M-step seeks an alignment that maximizes

$$\sum_i \hat{x}_i \log \Pr(q, M \mid x_i = 1, A^{(m)}) - C' = \sum_i \hat{x}_i \sum_j \log \frac{\Pr(y_j, Z_j \mid \tau_i^*)}{\Pr(y_j) \Pr(Z_j \mid \tau)}$$

$$= \sum_j \sum_i \hat{x}_i \log \frac{\Pr(y_j, Z_j \mid \tau_i^*)}{\Pr(y_j) \Pr(Z_j \mid \tau)} .$$

where j runs over the positions of the alignment $A^{(m)}$.

To find a local alignment $A^{(m)}$ maximizing a sum of per-position scores, one need only apply the standard Smith-Waterman algorithm [21] using an appropriate score function. In this case, we may define the score $\sigma(y, Z)$ between a residue y of q and a column Z of M as follows:

$$\sigma(y, Z) = \sum_i \hat{x}_i \log \frac{\Pr(y, Z \mid \tau_i^*)}{\Pr(y) \Pr(Z \mid \tau)} .$$

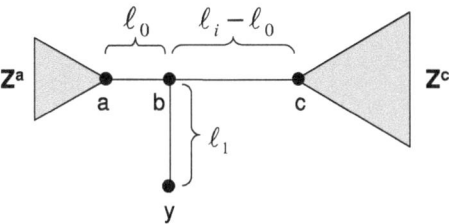

Fig. 2. The structure of tree τ_i^*, focusing on the branch e_i of τ, which lies between nodes a and c, and the new branch inserted at b, ending in the query residue y.

The scores $\sigma(y, Z)$ need not be recomputed from scratch for each cell of the dynamic programming matrix but can instead be computed for Z_j and all $y \in \Sigma$ as each column Z_j of M is first considered.

The EM algorithm requires a starting alignment $A^{(0)}$, which should be a reasonable guess at the final alignment. We derive $A^{(0)}$ by applying the general algorithm with a uniform distribution over the placement of q on τ, that is, with $\hat{x}_i = 1/(2n - 2)$.

4   Computation of Per-column Probabilities

We now turn to the problem of computing the probabilities $\Pr(y, Z \mid \tau_i^*)$, where y is a residue of q, Z a column of M, and τ_i^* an augmented tree topology. These probabilities or their logs are needed in both the E- and M-steps of EM, potentially for $|\Sigma|^n$ different values of Z and $2n-2$ different trees τ_i^*. If phylogenetically aware similarity search in large sequences is to be practically efficient, these probabilities must be computed as quickly as possible. Moreover, full specification of these probabilities reveals another piece of unknown information – a pair of branch lengths – that must be estimated from the data.

4.1   The Basic Computation

We wish to compute $\Pr(y, Z \mid \tau_i^*)$, where the topology τ_i^* appends the new branch leading to the query q from a branch e_i of τ with length ℓ_i. The postulated tree structure is shown in simplified form in Figure 2. Branch e_i divides τ into two subtrees whose roots are its endpoints. Let a and c be the roots of these two subtrees, and let Z^a and Z^c be the subsets of Z labeling their respective leaves. The new branch leading to the query residue y branches off e_i at some point; let b be the ancestral residue at this point. From this point to the end of Section 4.2, we focus on a particular choice of topology τ_i^* and so drop dependence on τ_i^* from our notation until Section 4.3.

Applying the conditional independence relations implied by the structure in the figure, we may derive

$$\Pr(y, Z) = \sum_{a,b,c} \Pr(y, Z, a, b, c)$$

$$= \sum_{a,b,c} \Pr(Z, a, c) \Pr(y, b \mid Z, a, c)$$

$$= \sum_{a,b,c} \Pr(Z^a, a) \Pr(Z^c, c \mid a) \Pr(y, b \mid a, c)$$

$$= \sum_{a,b,c} \Pr(Z^a \mid a) \Pr(a) \Pr(Z^c \mid c) \Pr(c \mid a) \Pr(y \mid b) \Pr(b \mid a, c) \ .$$

Applying Bayes' theorem to $\Pr(b \mid a, c)$ and performing a little algebra, we are left with

$$\Pr(y, Z) = \sum_{a,c} \Pr(Z^a \mid a) \Pr(Z^c \mid c) \sum_{b} \Pr(b) \Pr(y \mid b) \Pr(a \mid b) \Pr(c \mid b) \ .$$

The two terms in the outer sum, which depend on Z, can be computed given only the topology and branch lengths of τ. Inside the inner sum, we must apply our residue evolutionary model to the augmented tree. $\Pr(b) = \pi(b)$, the stationary probability of b. For the other three probabilities, we must introduce two new branch lengths, ℓ_0 and ℓ_1, as shown in Figure 2. ℓ_0 is the distance along e_i from a to the branch point b, while ℓ_1 is the length of the new branch leading to y.

Let $f_\ell(p \mid q)$ be the probability, according to the mutation rate matrix Q, that residue q mutates into residue p along a branch of length ℓ. Then we may write

$$\Pr(y, Z) = \sum_{a,c} \Pr(Z^a \mid a) \Pr(Z^c \mid c) \sum_{b} \pi(b) f_{\ell_1}(y \mid b) f_{\ell_0}(a \mid b) f_{\ell_i - \ell_0}(c \mid b) \ . \quad (5)$$

4.2   Estimating the Branch Lengths ℓ_0 and ℓ_1

The branch lengths ℓ_0 and ℓ_1 are unknown but must be supplied to compute $\Pr(y, Z)$. We therefore compute maximum-likelihood estimates for ℓ_0 and ℓ_1 given the current alignment A between q and M. That is, we choose these lengths to maximize

$$L(\ell_0, \ell_1 \mid A) = \prod_{(y,Z) \in A} \Pr(y, Z \mid \ell_0, \ell_1) \ ,$$

where each term of the product is given by Equation (5). Note that ℓ_1 is constrained to lie in the range $[0, \ell_i]$, while ℓ_0 is merely non-negative. As usual, the demands of numerical precision lead us to maximize the log of the likelihood L rather than L itself.

The optimal choice of branch lengths depends on the rate matrix Q. For all but the simplest Q, analytic determination of the optimal lengths appears difficult. We therefore maximize the log likelihood numerically using an implementation [22] of a bound-constrained quasi-Newton method. An important part of this process is the computation from Q of the transition probabilities $f_\ell(p \mid q)$

for a branch length ℓ and a pair of residues p, q. We use the fact [23] that $f_\ell(p \mid q)$ can be expressed as a sum

$$f_\ell(p \mid q) = \sum_{m=1}^{|\Sigma|} U_{qm} U_{mp}^{-1} e^{\lambda_m \ell} ,$$

where $Q = U \Lambda U^{-1}$ is the spectral decomposition of the rate matrix Q, with $\lambda_m = \Lambda[m, m]$ being its mth eigenvalue. The derivative $\frac{df_\ell(p|q)}{d\ell}$ is easily computed from this formula, which allows us to supply an analytic gradient $\nabla \cdot \log L(\ell_0, \ell_1 \mid A)$ to the optimizer.

To compute the initial alignment $A^{(0)}$ for EM, we arbitrarily set $\ell_0 = \ell_i/2$ and $\ell_1 = 0.1$ for all edges e_i of τ.

4.3 Caching to Accelerate Probability Computation

High-throughput use of phylogeny in sequence alignment requires numerous evaluations of the probability $\Pr(y, Z \mid \tau_i^*)$ for various augmented tree topologies τ_i^*. To speed these computations, we now consider how to partially precompute the necessary probabilities. Although the probabilities depend on branch lengths ℓ_0 and ℓ_1 for each τ_i^*, which must be recomputed in every E-step, some of the work necessary to compute these probabilities can be done either offline or once per update of the branch lengths.

In the expression for $\Pr(y, Z \mid \tau_i^*)$ in Equation (5), each term of the outer sum over a, c is divided into three components: $\Pr(Z^a \mid a)$, $\Pr(Z^c \mid c)$, and the inner summation over b. The latter sum depends on ℓ_0 and ℓ_1, but once these are known, it can be computed for all triples y, a, c in time $\Theta(|\Sigma|^4)$ per branch and stored in a table of size $|\Sigma|^3$. Each of the other two components, of the form $\Pr(Z^q \mid q)$, can be precomputed from the original tree τ and stored in a table of dimension $|\Sigma| \times |\Sigma|^{|Z^q|}$. These tables can be computed and stored offline because they do not depend on the query.

Once all three components have been precomputed, $\Pr(y, Z \mid \tau_i^*)$ requires only $\Theta(|\Sigma|^2)$ operations to sum over the ancestral residues a, c at either end of edge e_i. Hence, the score $\sigma(y, Z)$ can be computed in time $\Theta(n|\Sigma|^2)$. If we compute $\sigma(y, Z)$ for all y when the M-step first encounters a column Z of M, then the added cost to alignment is limited to $\Theta(n|\Sigma|^3)$ per column, regardless of the bandwidth used or the query size.

A potential problem with the above precomputation strategy is the large space requirement of the tables for the probabilities $\Pr(Z^q \mid q)$. The number of leaves Z^q behind an internal node q can be as large as $n - 1$, making the worst-case size of this table $\Theta(|\Sigma|^n)$ and the total space cost for all tables $O(n|\Sigma|^n)$. Moreover, this upper bound is tight because the n edges leading to leaves of τ each separate a leaf from a subtree with $n - 1$ leaves.

To address the space usage of precomputation for larger n, we now show how to reduce this space to $O(n|\Sigma|^{2n/3+2})$ while increasing the cost of computing $\Pr(y, Z)$ by only a factor of $|\Sigma|$. Consider the tree of Figure 3, rooted at an

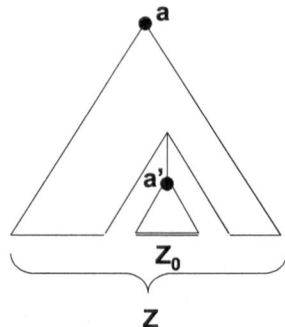

Fig. 3. Dividing a tree into two leaf sets Z_0 and $Z - Z_0$ around an internal node a'.

ancestral node a with a set Z of k leaves. Starting from a, walk down the tree, each time picking the side with more leaves, until we reach a node a' with at most $2k/3$ leaves below it. Let Z_0 be the set of leaves below a', and let $Z_1 = Z - Z_0$. The parent b of a' has more than $2k/3$ leaves below it, and a' roots the larger of b's two subtrees, so we have that $k/3 < |Z_0| \le 2k/3$, and hence that $k/3 \le |Z_1| < 2k/3$.

Observe that

$$\Pr(Z \mid a) = \sum_{a'} \Pr(a' \mid a) \Pr(Z_1 \mid a, a') \Pr(Z_0 \mid a') \ .$$

The three terms of this sum can be precomputed using tables of sizes $|\Sigma|^2$, $|\Sigma|^{|Z_1|+2}$, and $|\Sigma|^{|Z_0|+1}$, respectively. Moreover, we have that $|Z_0| \le 2k/3$ and $|Z_1| < 2k/3$, so the largest table size is at most $|\Sigma|^{2k/3+2}$. Given the three tables, the sum itself can be computed for any a and Z in time $\Theta(|\Sigma|)$.

Again, the largest table stored at any node of τ is for a subtree with $n - 1$ leaves, and this size is achieved for n nodes, so the total space cost is $O(n|\Sigma|^{2n/3+2})$. For certain tree topologies, the space cost may be as little as $O(n|\Sigma|^{n/2+2})$. More complex recursive applications of the above splitting procedure can trade off further exponential decreases in the size of stored tables for polynomial increases in online computation cost.

4.4   Missing Sequence Data

Short gaps in a multiple sequence alignment can be modeled as arising from a process of localized residue insertion and deletion, which can be incorporated into the mutation rate matrix Q. However, long-range genomic multiple alignments often contain large gaps caused by rearrangements or block insertions and deletions. Because these gaps do not arise from the local process modeled by the rate matrix, it seems wise to treat them separately. We therefore treat gaps in the sequences of M above a threshold length θ as missing data.

When aligning a query q to a region of M with missing sequences, we wish to consider only the non-missing subset of M and its induced subtree of τ in the alignment algorithm. In the limiting case that M contains only one non-missing sequence, the alignment problem reduces to simple pairwise sequence alignment. However, storing multiple induced subtrees, with all their scoring information, requires an infeasible amount of space. Instead, we treat missing positions uniformly by adding a *missing symbol* "*" to the list of symbols that can label the leaves of τ.

The defining property of the missing symbol is that $\Pr(*) = \Pr(* \mid a) = 1$. For the simplest rooted tree consisting of an ancestral node a and two leaves, it follows that

$$\Pr(*, y \mid a) = \Pr(y \mid a)$$
$$\Pr(*, * \mid a) = 1$$
$$\Pr(a \mid *, *) = \pi(a) \ .$$

For more complex trees, setting some subset of the leaves to the missing symbol effectively removes those species from computations on the tree. We may extend the precomputed tables for $\Pr(Z^q \mid q)$ to allow some subset of Z^q to be missing, which effectively increases the alphabet size for these tables by one. However, the missing symbol is not a full-fledged alphabet character and is not used other than to index the precomputed tables.

A similar approach to uniform treatment of missing residues is mentioned briefly in [24].

5   Results

We have implemented PhyLAT in the C++ language, using the OPT++ optimization library [22] to estimate parameters ℓ_0 and ℓ_1 for each edge in the E-step of the algorithm. In this section, we report initial performance tests using PhyLAT to align query sequences to orthologous regions in an alignment of fragments from eight vertebrate genomes.

5.1   Data Set and Parameterization

We tested PhyLAT on the "Zoo" set of vertebrate genomic sequences from the CFTR locus, which was obtained by Thomas et al. [3]. A multiple alignment of these sequences produced by MAVID [6] was obtained from NCBI, along with an induced tree labeled with branch lengths. From this alignment, we extracted a subalignment of eight organisms: chicken, mouse, rat, baboon, human, dog, cat, and cow. This subalignment spanned 4.4 million columns. From the same alignment, we extracted the individual sequences from pig and chimp, of lengths 1.15 and 1.45 megabases respectively after removing gaps. We first masked the low-complexity and repetitive elements in these two sequences using RepeatMasker [25]. We then sought high-scoring local alignments between these sequences and the multiple alignment.

We parameterized PhyLAT's evolutionary model for this test to match the observed properties of the multiple alignment. Gaps longer than $\theta = 21$ residues were first converted to missing data as described in Section 4.4. Stationary frequencies of the four nucleotides and the gap character were then inferred directly from their frequencies in the alignment. The remaining three rate parameters of the model were chosen to maximize the likelihood of the alignment given the tree, branch lengths, and stationary frequencies and were scaled to yield a total rate of one accepted mutation per unit of branch length.

Ungapped seed alignments between each query and the ancestral sequence α_M were used only if their score under the DNA-PAM-TT-50 scoring matrix corresponded to a Karlin-Altschul E-value of at most 10^{-5}. In gapped extension, we used a bandwidth $w = 101$ and gap penalties of 12 bits to open and 2 bits to extend, as suggested for pairwise alignment in [16].

5.2   Accuracy: Assessment of Tree Placement

Because the accuracy of the detailed alignment of each query sequence to the multiple alignment is difficult to assess, we sought alternate evidence of PhyLAT's accuracy that admitted easier assessment. Assuming that the original MAVID alignment of the Zoo data is broadly correct, a large majority of hits between the repeat-masked query q and the multiple alignment M should represent orthologous sequences, for which the species tree gives the correct placement of q on the tree τ for M. We therefore assessed how often PhyLAT's local alignments placed a query sequence on or near the expected edge of τ given the query's species.

For each of the two query species, we first identified those seed alignments likely to represent matches of the query to an orthologous region of the multiple alignment. Specifically, we sorted all seed alignments produced for each query by their starting positions in q, then retained only the subset of seed alignments whose starting positions in M formed a longest increasing subsequence. This procedure eliminated 5.1% of seed alignments to pig and 10.8% of seed alignments to chimp, leaving 1348 and 2705 alignments, respectively. Each seed alignment was then subjected to gapped extension against M, yielding 1132 and 2198 unique gapped alignments for pig and chimp, respectively.

Figure 4 shows the tree for M. The correct edges for the pig and chimp sequences are respectively those leading to the leaves labeled "cow" and "human." Among PhyLAT's alignments, 814/1132 (71.9%) for pig and 2068/2198 (94.1%) for chimp placed their queries on the correct tree edge for their species. Moreover, define an *almost correct* edge to be one that is either correct or shares an endpoint with the correct edge. By this definition, 1004/1132 (88.7%) alignments for pig and 2183/2198 (99.3%) alignments for chimp placed their queries on an almost correct edge.

It should be noted that, for this alignment M, PhyLAT typically had many fewer than eight sequences to help it place the query on the tree. The average number of species present at any point in M is only 2.5, so the majority of species may be expected to be absent at any given locus. However, not all species are missing equally often, and not all are equally helpful in aligning queries. In particular, if we consider the nearest neighbors of the two query species (cow

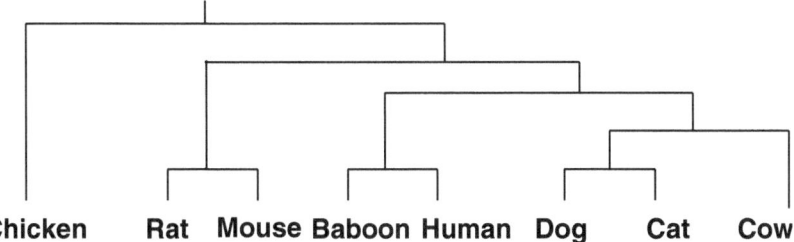

Fig. 4. Topology of the multiple alignment used for validation of PhyLAT. Branch lengths are not to scale.

for pig, human for chimp), we find that human sequence is present at 42% of positions, while cow is present at only 34%. This difference in coverage, as well as the smaller evolutionary distance for chimp-human vs pig-cow, likely explains the observed difference in performance on the two query species.

5.3 Efficiency

We tested PhyLAT on a 2.5 GHz Intel Pentium 4 workstation. The times to perform gapped extension for the seed alignments inspected for pig and chimp were 472s and 1290s, respectively. Time to generate the seed alignments was comparatively negligible. Running time was divided roughly 3:1 between the M-step and the E-step computations. For the M-step, about 40% of time was spent executing Smith-Waterman, while the remaining 60% was spent match computing scores for each column of M used in alignment. For the E-step, almost all time was spent in evaluating augmented tree likelihoods (and their derivatives) for various choices of e_i, ℓ_0, and ℓ_1 in the optimizer.

We note that most of the time spent computing probabilities was consumed performing table lookups and multiplications to reconstruct them from the precomputed intermediates described in Section 4.3.

6 Conclusions and Open Problems

We have described a probabilistic model and algorithm for obtaining and scoring local alignments between a query sequence q and a genomic multiple alignment M. The scoring scheme explicitly incorporates information from a phylogenetic tree on M, while the alignment algorithm is motivated by finding the most likely alignment given uncertainty about the position of q on this tree. We address the need for precomputation to accelerate the algorithm while maintaining reasonable storage requirements. An initial implementation of our method in the PhyLAT search tool shows promise for accurately reconstructing the relationship of q to M.

Several challenges remain to make PhyLAT robust enough for production use. A first important issue is the treatment of gaps and, more generally, the evolutionary model associated with τ. Our current treatment of gaps was chosen for computational convenience but is limited in its ability to model gaps of different lengths accurately. Placing gaps correctly in a multiple alignment is still not a solved problem, so a more accurate gap model may be substantially more costly in practice. We can improve our scoring of gaps *and* other columns somewhat by using a context-dependent rate matrix as described in [24]. Such a model would necessitate considerably larger cached tables, which would require aggressive application of table splitting.

The multiple alignment M, while it likely has a single consistent tree topology τ, need not have a single globally correct set of branch lengths. This concern could be addressed by segmenting M into regions with relatively constant branch lengths. The number of different regions must be limited to avoid untenable storage costs for precomputed probabilities, but at least 5-10 types of region, each with different branch lengths, would likely be feasible for alignments of about ten species. Additional complexities arise if a single alignment spans multiple regions.

Although we have generated scores for our alignments, we have not yet devised a mechanism for measuring their statistical significance. Because the alignment algorithm used is essentially Smith-Waterman, we may hope that Karlin-Altschul theory [19] will enable us to convert ungapped alignment scores into more customary E-values, and that gapped extension will follow a similar extreme value distribution that can be estimated empirically.

Finally, an important open problem is how to proceed when the query is not a single sequence but is itself an alignment of two or more sequences. The number of possible augmented tree topologies undergoes combinatorial explosion with multiple query sequences, so explicitly enumerating them is likely infeasible. More efficient methods will be needed to find good hypotheses for combining complex tree topologies.

References

1. Cliften, P., Sudarsanam, P., Desikan, A., Fulton, L., Fulton, B., Majors, J., Waterston, R., Cohen, B.A., Johnston, M.: Finding functional features in saccharomyces genomes by phylogenetic footprinting. Science **301** (2003) 71–6
2. Bahl, A., Brunk, B., Crabtree, J., Fraunhoz, M.J., et al.: PlasmoDB: the *Plasmodium* genome resource. Nucleic Acids Research **31** (2003) 212–5
3. Thomas, J.W., Touchman, J.W., Blakesley, R.W., Bouffard, G.G., et al.: Comparative analyses of multi-species sequences from targeted genomic regions. Nature **424** (2003) 788–93
4. Schwartz, S., Zhang, Z., Frazer, K.A., Smit, A.F., et al.: PipMaker – a web server for aligning two genomic DNA sequences. Genome Research **10** (2000) 577–86
5. Höhl, M., Kurtz, S., Ohlebusch, E.: Efficient multiple genome alignment. Bioinformatics **18** (2002) S312–20
6. Bray, N., Dubchak, I., Pachter, L.: AVID: a global alignment program. Genome Research **13** (2003) 97–102

7. Brudno, M., Do, C., Cooker, G., Kim, M.F., Davydov, E., Green, E.D., Sidow, A., Batzoglou, S.: LAGAN and Multi-LAGAN: efficient tools for large-scale multiple alignment of genomic DNA. Genome Research **13** (2003) 721–31
8. Siepel, A., Haussler, D.: Computational identification of evolutionarily conserved exons. In: Proceedings of the Eighth Annual International Conference on Computational Molecular Biology (RECOMB04), San Diego, CA (2004) 177–86
9. McAuliffe, J.D., Pachter, L., Jordan, M.I.: Multiple-sequence functional annotation and the generalized hidden Markov phylogeny. Bioinformatics **20** (2004) 1850–60
10. Altschul, S.F., Gish, W.: Local alignment statistics. Methods: a Companion to Methods in Enzymology **266** (1996) 460–80
11. Altschul, S.F., Madden, T.L., Schäffer, A.A., Zhang, J., Zhang, Z., Miller, W., Lipman, D.J.: Gapped BLAST and PSI-BLAST: a new generation of protein database search programs. Nucleic Acids Research **25** (1997) 3389–402
12. Yona, G., Levitt, M.: A unified sequence-structure classificatin of proteins: combining sequence and structure in a map of protein space. In: Proceedings of the Fourth Annual International Conference on Computational Molecular Biology (RECOMB00), Tokyo, Japan (2000) 308–17
13. Wang, T., Stormo, G.D.: Combining phylogenetic data with co-regulated genes to identify regulatory motifs. Bioinformatics **19** (2003) 2369–80
14. Tamura, K., Nei, M.: Estimation of the number of nucleotide substitutions in the control region of mitochondrial DNA in humans and chimpanzees. Molecular Biology and Evolution **10** (1993) 512–26
15. McGuire, G., Denham, M.C., Balding, D.J.: Models of sequence evolution for DNA sequences containing gaps. Molecular Biology and Evolution **18** (2001) 481–90
16. Durbin, R., Eddy, S., Krogh, A., Mitchison, G.: Biological Sequence Analysis. Cambridge University Press, New York (1998)
17. Buhler, J., Keich, U., Sun, Y.: Designing seeds for similarity search in genomic DNA. In: Proceedings of the Seventh Annual International Conference on Computational Molecular Biology (RECOMB '03), Berlin, Germany (2003) 67–75
18. States, D.J., Gish, W., Altschul, S.F.: Improved sensitivity of nucleic acid database searches using application-specific scoring matrices. Methods: a Companion to Methods in Enzymology **3** (1991) 66–70
19. Karlin, S., Altschul, S.F.: Methods for assessing the statistical significance of molecular sequence features by using general scoring schemes. PNAS **87** (1990) 2264–8
20. Dempster, A.P., Laird, N.M., Rubin, D.B.: Maximum likelihood from incomplete data via the EM algorithm. Journal of the Royal Statistical Society B **39** (1977) 1–38
21. Smith, T.F., Waterman, M.S.: Identification of common molecular subsequences. Journal of Molecular Biology **147** (1981) 195–97
22. Meza, J.C., Hough, P.D., Williams, P.J.: Opt++ optimization library 2.1r3 (2004) http://csmr.ca.sandia.gov/projects/opt++.
23. Strimmer, K., von Haeseler, A.: Nucleotide substitution models. In Salemi, M., Vandamme, A.M., eds.: The Phylogenetic Handbook. Cambridge University Press, New York (2003)
24. Siepel, A., Haussler, D.: Phylogenetic estimation of context-dependent substitution rates by maximum likelihood. Molecular Biology and Evolution **21** (2004) 468–88
25. Smit, A.F., Green, P.: Repeatmasker (1999) http://ftp.genome.washington.edu/RM/RepeatMasker.html.

Multiple Genome Alignment
by Clustering Pairwise Matches

Jeong-Hyeon Choi[1,3]*, Kwangmin Choi[1], Hwan-Gue Cho[3], and Sun Kim[1,2]*

[1] School of Informatics, Indiana University, IN 47408, USA
{jeochoi,kwchoi,sunkim}@bio.informatics.indiana.edu
[2] Center for Genomics and Bioinformatics, Indiana University, IN 47405, USA
[3] Department of Computer Science and Engineering,
Pusan National University, Korea
{jhchoi,hgcho}@pusan.ac.kr

Abstract. We have developed a multiple genome alignment algorithm
by using a sequence clustering algorithm to combine local pairwise
genome sequence matches produced by pairwise genome alignments, e.g,
BLASTZ. Sequence clustering algorithms often generate clusters of sequences such that there exists a common shared region among all sequences in each cluster. To use a sequence clustering algorithm for genome
alignment, it is necessary to handle numerous local alignments between
a pair of genomes. We propose a multiple genome alignment method
that converts the multiple genome alignment problem to the sequence
clustering problem. This method does not need to make a guide tree
to determine the order of multiple alignment, and it accurately detects
multiple homologous regions. As a result, our multiple genome alignment algorithm performs competitively over existing algorithms. This is
shown using an experiment which compares the performance of TBA,
MultiPipMaker (MPM) and our algorithm in aligning 12 groups of 56
microbial genomes and by evaluating the number of common COGs detected.

1 Introduction

Recent advances in both sequencing technology and algorithm development for
genome sequence software have made it possible to determine the sequence of a
whole genome. As a consequence, the number of completely sequenced genomes
is increasing rapidly. However, algorithm development for genome annotation
has been relatively slow, and the annotation of completely sequenced genomes
inevitably depends on human expert knowledge. The most effective method to
understand genome content is to compare multiple genomes, especially when
they are close enough to share common subsequences. One important computational method is to align whole genome. Aligning whole genomes is useful in
several ways. For example, sequencing a genome can be much easier and more
accurate by aligning its *contigs* to completely sequenced genomes that are close

---

* Corresponding author.

J. Lagergren (Ed.): RECOMB 2004 Ws on Comparative Genomics, LNBI 3388, pp. 30–41, 2005.
© Springer-Verlag Berlin Heidelberg 2005

to the one being sequenced. Another example of use of whole genome alignment is to identify conserved regions among multiple genomes which include not only common genes but regulatory regions and non-coding RNA sequences. A recent paper demonstrated that intergenic functional regions in multiple Yeast strains can be detected by comparing the whole genomic sequences [1]. Work on comparing human and mouse sequences has also demonstrated the possibility of predicting functions and structures of human genes by genome alignment methods, e.g., [2, 3].

The dynamic programming algorithms, such as Needleman-Wunsch [4] and Smith-Waterman [5], can be extended to align optimal alignment of multiple sequences, e.g., MSA [6]. However, computing optimal alignments for long genomic sequences is not practical in terms of computation time and memory requirement; assuming N sequences of equal length L, time complexity is $O(2^N L^N)$ and the space complexity is $O(L^N)$. Thus heuristic approaches, such as progressive and iterative alignments, are widely used.

The progressive alignment (1) aligns all pairs of sequences by a fast anchor-based alignment approach or a slow full dynamic programming method like Needleman-Wunsch algorithm, (2) produces a guide tree via distance matrix using the pairwise alignment scores, and (3) aligns the sequences sequentially guided by the phylogenetic relationships indicated by the tree. The major problem with progressive alignment programs such as CLUSTALW [7] and PILEUP in GCG Wisconsin Package is that the order of progressive alignment is determined and fixed by the initial pairwise alignments. To overcome this problem, some iterative alignment methods make initial alignments for groups of sequences and then revise the alignments to compute a more accurate result by various approaches [8–11].

Unlike the progressive and iterative alignment strategy, chaining alignment strategy aligns multiple sequences without making guide tree. This method is based on chaining all pairwise local alignments or multiple local alignments. MGA [12] first finds multiMEMs, maximal exact matches occurring in multiple genomes, as anchors and then calculates the optimal chain for multiMEMs by using graph or range tree. DIALIGN [13] aligns pairs of sequences to locate aligned regions that do not include gaps, that is, continuous diagonals in a dot matrix plot. A consistent collection of weighted diagonals is then computed, and diagonals with maximal sum of weights are generated as alignments.

As more whole genome sequences become available, there has been growing need for computational methods for aligning multiple whole genomes. MLAGAN [14] aligns genomic sequences in progressive alignment phases with LAGAN and optimal iterative improvement phases. MAVID [15] uses a progressive alignment approach to incorporate the following ideas: maximum likelihood inference of ancestral sequences, automatic guide tree construction, protein-based anchoring of ab-initio gene predictions, and constraints derived from a global homology map of the sequences. TBA [16] builds a threaded blockset under the assumption that all matching segments occur in the same order and orientation in given sequences; inversions and duplications are not addressed. TBA is shown to generate very high quality multiple genome alignments that verified with their rigorous column-by-column comparing evaluation method. MultiPipMaker [17] aligns a reference sequence individually with each secondary sequence, prepares a crude multiple alignment from the pairwise alignments, removes overlaps in the local pairwise alignments, and refines the crude multiple alignment to generate a true multiple alignment using rigorously defined multiple alignment scores.

1.1   Motivation

As we discussed, a heuristic alignment approach is the most widely used technique for multiple sequence alignment. However, the following reasons make it difficult to use heuristic alignment strategy for the multiple genome alignment problem.

1. It is hard to utilize the guide tree since there are many local regions where their phylogenetic relationships are different from those of the "entire" genome. Even duplications of a gene within a genome have their own phylogenetic relationship.
2. The greedy progressive alignment works due to the guide tree. Given that the utilization of a guide tree is not trivial for genome alignment as discussed above, the greedy progressive alignment strategy should be avoided.
3. The iterative alignment requires generation of profiles – alignment of multiple sequences – while combining pairwise matches. Generating and aligning profiles for long genome sequences is not practical.

An alternative to the progressive alignment method is to compute subsequences common to all genomes being aligned and chain them together to generate multiple sequence alignment. For example, MGA [12] computes multiMEMs common in all genomes being aligned. MEMs in multiple genomes are naturally short in length and thus it is necessary to chain them to generate multiple sequence alignments. Although MGA was successful in generating "global" alignments of closely related genomes, it is not effective to compute alignments for relatively distant genomes. One way to circumvent the difficulty is to use directly pairwise genome alignments which are already extended matches between a genome pair. Thus some algorithms such as TBA [16] and MultiPipMaker [17] are designed to combine pairwise matches for the multiple genome alignment problem.

We have developed a multiple genome alignment algorithm by using a sequence clustering algorithm [18] to combine pairwise matches. Sequence clustering algorithms generate clusters of sequences that are candidates for sequence families. These clusters are often generated to have common shared regions among all sequences in each cluster, and these shared regions are used to predict sequence domains. We argue that sequence clustering algorithms can be used to generate multiple genome alignments by combining pairwise genome matches since sequence clustering algorithms often require all pairwise alignments of the input sequence set and combine pairwise matches to generate sequence clusters. The genome alignment problem, however, is different from the sequence clustering problem in terms of the number of input sequences and the number of

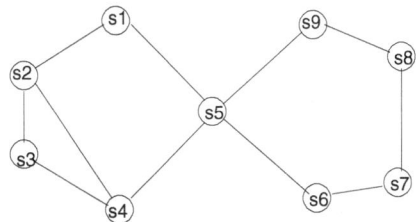

Fig. 1. Biconnected components and articulation point. There are two biconnected components, $\{s_1, s_2, s_3, s_4, s_5\}$ and $\{s_5, s_6, s_7, s_8, s_9\}$. The vertex s_5 is an articulation point since removing the vertex results in splitting the graph.

alignments between sequences. The number of genomes to be aligned is much smaller than that of sequences to be clustered. In addition, there are numerous local alignments between a pair of genomes while one or only a few aligned regions exist between a pair of (protein) sequences. The basic idea of this paper is to transform the genome alignment problem to the sequence clustering problem, which will be tackled by our sequence clustering algorithm BAG [18]. Thus we explain briefly how BAG works in the next section and we describe how to transform the genome alignment problem to the sequence clustering problem by generating subsequences in Section 2.2.

2   A Sequence Clustering Algorithm and the Generation of Input Data from Pairwise Matches

2.1   BAG Sequence Clustering Algorithm

Let us first review some definition of graph. A *connected component* of a graph G is a subgraph where any two vertices in the subgraph are reachable from each other. An *articulation point* of G is a vertex whose removal disconnects G. For example, in Fig. 1, the removal of a vertex s_5 disconnects G. A *biconnected graph* is a graph where there are at least two disjoint edge paths for any pair of vertices. A *biconnected component* of G is a maximal biconnected subgraph. In Fig. 1, a subgraph G_1 induced by vertices $\{s_2, s_3, s_4\}$ is a biconnected graph but it is not a biconnected component since another subgraph G_2 induced by vertices $\{s_1, s_2, s_3, s_4, s_5\}$ is biconnected and G_1 is a subgraph of G_2. There are two biconnected components, $\{s_1, s_2, s_3, s_4, s_5\}$ and $\{s_5, s_6, s_7, s_8, s_9\}$ of G in Fig. 1.

For a given set of (protein) sequences $\{s_1, s_2, ..., s_n\}$, a weighted graph G can be built from all pairwise comparison results by using FASTA and BLAST. A node is created for each sequence s_i and an edge between two sequences, s_i and s_j, is created when the pairwise alignment score of s_i and s_j is more significant than a preset threshold. The alignment score is associated with the edge as weight so that clusters can be refined while increasing cutoff for edges. Then our sequence clustering algorithm BAG will be used for multiple genome alignment later in this paper. BAG explicitly uses two graph properties: biconnected components and articulation points (see Fig. 1). We argue that a biconnected component (*BCC* in short) is a candidate for a family of sequences because biconnected graph requires at least two disjoint edge paths between *every pair of*

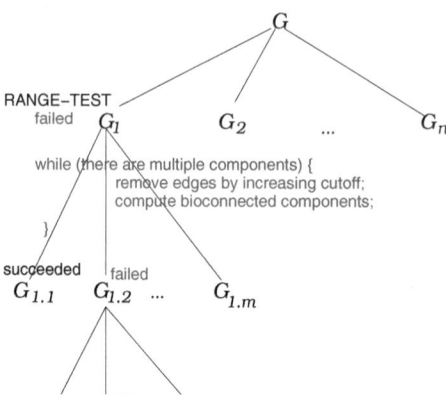

Fig. 2. Illustration of the splitting step. For any biconnected component that failed for RANGE-TEST, a cutoff score is increased until it can be split into multiple biconnected components.

nodes in the graph and requires a much stronger condition than single linkage. We also argue that an articulation point is a candidate for a multidomain protein since it is the only vertex that connects two or more biconnected components, i.e., multiple families. According to the graph properties mentioned previously, we named our algorithm as BAG which stands for Biconnected components and Articulation point based Grouping of sequences.

For a given set of sequences $\{s_1, s_2, ..., s_n\}$, BAG recursively computes the biconnected components with an increased cutoff score at each iteration. Our algorithm BAG works as follows:

1. Build a weighted graph G from all pairwise comparison results where a node is created for each sequence s_i and an edge between two sequences, s_i and s_j, is created when the pairwise alignment score of s_i and s_j is more significant than a preset threshold. The alignment score is associated with the edge as weight so that clusters can be refined while increasing cutoff for edges.
2. Select a cutoff score T_c, remove edges with weight smaller than T_c, and generate biconnected components, $G_1, G_2, ..., G_n$, with a set of articulation points $\{a_1, a_2, ..., a_m\}$. See [18] for the detail how to select this cutoff score.
3. Perform RANGE-TEST for each biconnected component G_i to determine whether all sequences in G_i have common shared regions. If G_i fails RANGE-TEST, split G_i iteratively and recursively into multiple clusters $\{G_{i_1}, ..., G_{i_l}\}$ until every biconnected component G_{i_j} passes the RANGE-TEST as increasing the cutoff score by δ [1]. Note that increasing the cutoff score will remove some edges and the resulting graph can be split into multiple biconnected components at a certain cutoff. This step is illustrated in Fig. 2.

---

[1] While increasing the cutoff score, we have several competing cutoff values that could generate different clustering result. In the past, we selected the first cutoff that splits G_i into multiple clusters. We recently developed a better method, called *Cluster Utility* to select the best one among different cutoff choices. See [19] for more detail.

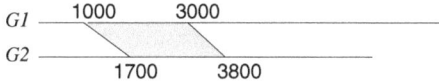

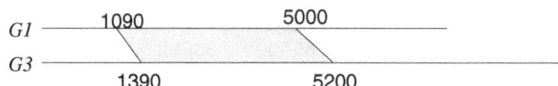

Fig. 3. Difficulty in generating subsequences from pairwise alignments. In G_1, there are two alignments at similar positions, one with G_2 and another with G_3. One straightforward way to name subsequences is to assign a name with starting positions of alignment in each genome. For G_1, we would generate two subsequences G_1:1000 and G_1:1090. However, these two subsequences correspond to the same homologous region and it is difficult to match them. In addition, this result in generating numerous subsequences when there are multiple genomes; in the worst case, the number of subsequences of a genome can be the length of the genome.

4. After iterative splitting is done (step 2 and 3), clusters are considered for merging. Each articulation point is tested for having common shared regions with its neighbors in different clusters; we call this the AP-TEST. Then a hypergraph is built as follows. Clusters from the previous step become vertices and articulation points that fail AP-TEST become edges in the hypergraph. A set of biconnected components of the hypergraph is iteratively merged into one until there is no candidate component for further merging as relaxing the cutoff score T_c.

2.2 Generation of the Input Data to BAG

BAG is originally designed to handle protein sequences which are typically 1,000 amino acid characters or less in length. So it cannot directly handle all pairwise alignments of genomes since there are numerous edges (pairwise matches) between a pair of nodes (genomes). This issue is handled by generating subsequences with their own identifiers. However, the generation of subsequences is not straightforward as illustrated in Fig. 3 where two different identifiers are generated for a single homologous region. Our approach is to convert local alignments to subsequence identifiers which start at one of evenly spaced break positions.

Let a break position be denoted by b and a local alignment α be represented by (s_i, e_i, s_j, e_j) where s_i and e_i (s_j and e_j) are the starting and end positions of α in genome G_i (G_j). Then α generates the subsequence identifiers $(G_i{:}p_ik, \ G_j{:}p_jk)$, $(G_i{:}p_i(k+1), \ G_j{:}p_j(k+1)),\ldots,$ $(G_i{:}q_i, \ G_j{:}q_j)$ where $p_i = b\lfloor s_i/b \rfloor$, $p_j = b\lfloor s_j/b \rfloor$, $q_i = b\lfloor e_i/b \rfloor$, $q_j = b\lfloor e_j/b \rfloor$. Table 1 shows the BAG input for local alignment regions (497087, 499555, 698686, 701075) for NC_000908 vs. NC_000912 pair, (497227, 498276, 2682403, 2683463) for NC_000908 vs. NC_000913 pair, and (698823, 699860, 2682400, 2683452) for NC_000912 vs. NC_000913 pair. As discussed, the alignments start at slight different positions,

Table 1. The bag input for local alignments, (497087, 499555, 698686, 701075) for NC_000908 vs. NC_000912 pair, (497227, 498276, 2682403, 2683463) for NC_000908 vs. NC_000913 pair, and (698823, 699860, 2682400, 2683452) for NC_000912 vs. NC_000913 pair to detect common COG0112.

seq1	seq2	score	pos1	pos2
NC_000908:497000	NC_000912:698000	60562	87,2555	686,3075
NC_000908:498000	NC_000912:699000	60562	0,1555	0,2075
NC_000908:499000	NC_000912:700000	60562	0,555	0,1075
NC_000908:497000	NC_000913:2682000	19888	227,1276	403,1463
NC_000908:498000	NC_000913:2683000	19888	0,276	0,463
NC_000912:698000	NC_000913:2682000	22113	823,1860	400,1452
NC_000912:699000	NC_000913:2683000	22113	0,860	0,452

for example, positions 497087 and 497227 for NC_000908, which is not trivial to handle given that there are numerous local matches for whole genomes. By generating subsequences that start at a 1000 bp break position ($b = 1000$), the two different starting positions, 497087 and 497227 for NC_000908, have the same identifier, NC_000908:497000, with their aligned positions adjusted as shown in Table 1. Now BAG can combine matching regions using subsequence identifiers into two clusters, {NC_000908:497000, NC_000912:698000, NC_000913:2682000} and {NC_000908:498000, NC_000912:699000, NC_000913:2683000}.

3   Strategy for Combining Pairwise Matches

Let us summarize the whole procedure for combining pairwise matches in order to generate multiple sequence alignments. For a set of genomes, $\{G_1, G_2, ..., G_n\}$,

1. Compute pairwise alignments using BLASTZ for each pair of genomes, G_i and G_j, $1 \le i < j \le n$.
2. Generate subsequence identifiers at each breakpoint as described in Section 2.2 so that all pairwise genome data can be used for BAG. This is necessary since each pairwise alignment for a local homologous region can start at different position. For example, consider three local pairwise matches for a gene a_1 in Fig. 4, $a_1 - a_2$, $a_1 - b_1$, and $a_1 - c_1$. Positions for a_1 in three pairwise matches start at slightly different positions in most cases and generation of subsequence identifiers can handle this easily (see Table 1 for a real example)
3. Once subsequence identifiers are generated, BAG enumerates biconnected components with a common shared region. For example, all five genes in Fig. 4 form a biconnected component, but they do not share a common shared region (we assume this since two different families are clustered together). Increasing a cutoff to 1500 eliminates an edge between c_1 and c_2 from the graph and recomputation will successfully separate the current graph into two biconnected components.

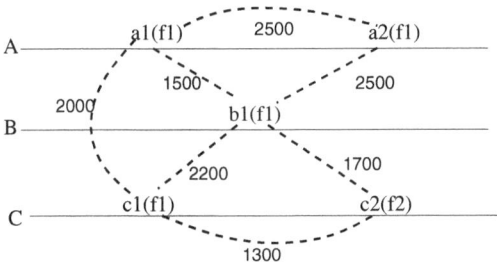

A, B, C: genomes
a1, a2, ... : genes in genome A
f1, f2, ... : gene family names
a1- $\frac{2500}{}$ - a2 : a match between a1 and a2 with a score 2500

Fig. 4. Co-occurrences of genes in three genomes.

4. Among final biconnected components, i.e., those that successfully passed RANGE-TEST, we select ones that cover at least k genomes. We call this *support value*. By default, k is the number of genomes being aligned. For the example in Fig. 4, $k = 3$. There are two biconnected components, $\{a_1, a_2, b_1, c_1\}$ and $\{b_1, c_2\}$, but $\{b_1, c_2\}$ is not considered as an alignment since SUPPORT($\{b_1, c_2\}$) < k.

4   Experimental Results

To evaluate the performance of our method, we have used completely sequenced microbial genomes published in NCBI, grouped them into sets of genomes based on taxonomy, and selected 12 groups as shown in Table 2.

The multiple alignment programs can be evaluated in terms of the number of common COGs detected by alignment results; the common COGs are extracted from protein table (PTT) files downloaded from GenBank at NCBI. In this experiment, "common" COGs are those that are present in "all" genomes being aligned. This criterion excludes any COGs present only in a proper subset of genomes being aligned. As an alternative to our evaluation method, it is worth noting the rigorous column-by-column comparing evaluation method that used by TBA. Our method confirms only whether common homologous regions can be detected or not, while the TBA method try to evaluate how accurate the alignment result is. In the current implementation of our algorithm, focus is just on how accurately pairwise alignments can be combined and the final alignments are generated using existing multiple sequence alignment algorithms such as CLUSTALW [7].

Given the goal of detecting common COGs present in all genomes, we can evaluate the performance of multiple genome alignment algorithms in two ways. The first evaluation criterion is that a common COG is detected only when the region aligned by the algorithm is present in all genomes. The second criterion is

Table 2. The 12 groups selected from microbial genomes for the experiments.

Actinobacteridae (5):	NC_003450, NC_002677, NC_002755, NC_000962, NC_004572
Alphaproteobacteria (5):	NC_004463, NC_003317, NC_002678, NC_003103, NC_000963
Bacillales (4):	NC_002570, NC_000964, NC_003212, NC_002745
Betaproteobacteria (3):	NC_003295, NC_003112, NC_003116
delta-epsilon (3):	NC_000915, NC_000921, NC_002163
Ecoli (3):	NC_000913, NC_002655, NC_002695
Enterobacteriaceae (4):	NC_000913, NC_003197, NC_003143, NC_002528
Euryarchaeota (10):	NC_002607, NC_000917, NC_000909, NC_003551, NC_003552, NC_000916, NC_002578, NC_002689, NC_000868, NC_000961
Firmicutes (7):	NC_003030, NC_002570, NC_000964, NC_003212, NC_002745, NC_003028, NC_002737
Mollicutes (5):	NC_000908, NC_004432, NC_000912, NC_002771, NC_002162
Mycoplasma (4):	NC_000908, NC_004432, NC_000912, NC_002771
Thermoprotei (3):	NC_000854, NC_003364, NC_002754

that a common COG is detected if the region aligned by the algorithm is present at least in the half of the all genomes, i.e, the support value is more than half of the number of genomes. The first criterion will be called as ALL and the second as HALF.

We compared performances of TBA, MultiPipMaker (MPM), and different versions of our algorithm BAG, BAG-BCC (Fw), BAG-BCC (Fw+Bw), BAG-CC (Fw), and BAG-CC (Fw+Bw); Fw means the case that only forward local alignments are used as TBA works, Fw+Bw means the case that both forward and backward local alignments are used, CC means that connectedness is used for clustering criterion, and BCC means that biconnectedness is used for clustering criterion. With the ALL criterion, BAG (Fw+Bw) and MPM consistently outperformed TBA. This is because TBA aligned forward strands only. To make comparison with TBA fair, the alignment result of BAG (Fw) is shown in the table. BAG (Fw) outperformed TBA except for *Actinobacteridae*, *delta-epsilon*, and *Enterobacteriaceae*. With the relaxed HALF criterion, both BAG and TBA are competitive. Detailed summary of the alignment results are given in Table 3. In the tables, the second column represents the number of common COGs among all genomes in each group, the third and the seventh columns represent the number of common COGs detected by TBA with ALL and HALF condition respectively, the fourth and ninth columns by MPM, the fifth and the tenth columns by BAG-BCC (Fw), and the sixth and the eleventh columns by BAG-BCC (Fw+Bw). TBA could not align *Ecoli* for a unknown system related reason. Note that BCC (biconnectedness) produced better results than CC (single linkage), which demonstrates clearly the effectiveness of biconnectedness though we omit it for page limit. It is also worth looking at how many detected alignments

Table 3. The alignment results of TBA, MPM, and BAG in terms of the number of common COGs detected with the ALL and HALF criterion.

Group	#COGs	ALL				HALF			
		TBA	MPM	BAG		TBA	MPM	BAG	
				Fw	Fw+Bw			Fw	Fw+Bw
Actinobacteridae	447	11	47	9	66	450	535	227	671
Alphaproteobacteria	468	0	17	1	62	17	357	87	487
Bacillales	935	85	192	91	398	756	1032	330	884
Betaproteobacteria	979	0	255	7	102	704	633	18	150
delta-epsilon	759	52	237	9	111	755	807	12	128
Ecoli	1972		1897	1004	1487		2006	1032	1555
Enterobacteriaceae	520	29	117	20	210	1694	1801	389	1123
Euryarchaeota	330	0	1	0	4	0	90	11	177
Firmicutes	573	0	12	2	39	87	300	189	514
Mollicutes	258	1	21	1	37	27	231	48	200
Mycoplasma	291	2	40	2	49	297	356	16	155
Thermoprotei	671	0	74	5	175	37	347	6	211

Table 4. The alignment results of TBA, MultiPipMaker (MPM) and BAG in terms of the number of clusters that detect common COGs (with the ALL criterion) and the number of clusters generated by the algorithm. The number pair n/m denote that m alignments were generated by the corresponding algorithm and, among them, n alignments correspond to the common COGs. Note that the number of clusters are significantly smaller than the number of COGs detected in Table 3 since an alignment covers multiple common COGs that appear in tandem.

Group	TBA	MPM	BAG			
			BCC		CC	
			Fw	Fw+Bw	Fw	Fw+Bw
Actinobacteridae	28 / 1731	45 / 3010	13 / 361	82 / 1632	1 / 70	9 / 20
Alphaproteobacteria	0 / 871	12 / 14426	2 / 130	68 / 1413	0 / 107	2 / 121
Bacillales	99 / 1514	169 / 6368	115 / 455	480 / 1411	42 / 211	46 / 176
Betaproteobacteria	0 / 328	244 / 1604	8 / 26	117 / 182	8 / 30	27 / 41
delta-epsilon	20 / 250	190 / 1433	11 / 12	94 / 127	12 / 13	36 / 44
Ecoli	/	379 / 1303	1043 / 1060	1957 / 2039	988 / 992	648 / 676
Enterobacteriaceae	40 / 1453	83 / 5489	26 / 523	231 / 1626	3 / 325	35 / 360
Euryarchaeota	0 / 736	1 / 3954	0 / 240	8 / 1264	0 / 152	0 / 63
Firmicutes	0 / 2000	20 / 10769	3 / 720	64 / 1985	2 / 294	1 / 147
Mollicutes	2 / 443	20 / 1311	2 / 61	51 / 343	1 / 47	6 / 33
Mycoplasma	2 / 298	36 / 1025	3 / 36	61 / 240	3 / 31	9 / 31
Thermoprotei	0 / 51	86 / 1414	9 / 12	274 / 338	9 / 11	112 / 136

correspond to the common COGs. Table 4 shows the result with the ALL criterion. In general, higher ratios of alignments by BAG correspond to common COGs than those by TBA and MPM. However, this does not necessarily imply that BAG generates more accurate alignment results since there are many regions shared among genomes that are outside common COGs. The primary role of genome alignment algorithms is probably to generate common shared regions in multiple genomes so that these regions can be investigated further biologically or computationally, thus our experiment suggests that BAG is complementary to existing algorithms such as TBA and MPM.

5   Conclusion

In this paper, we proposed a new strategy for multiple genome alignment. It uses a sequence clustering algorithm, BAG, to combine pairwise alignments. Our strategy accurately detected multiple homologous regions and performed competitively over existing algorithms as shown in the experiments with 12 groups of 56 microbial genomes in terms of the number of common COGs detected. This evaluation method is not sufficient in a sense because COGs are protein coding genes and there are many interesting regions common in multiple genomes such as non-coding RNAs and intergenic regions. Alignments of these regions together with biological interpretation will be reported in a forthcoming paper. However, the most primary use of genome alignment tools is to detect coding regions as mentioned in [20], and we believe that our evaluation method using common COGs is effective, especially for microbial genomes where intergenic regions are quite small.

Acknowledgments

This work partially supported by INGEN (Indiana Genomics Initiatives), NSF CAREER Award DBI-0237901, and KOSEF (Korea Science and Engineering Foundation) F01-2004-000-10016-0. We thank Haixu Tang of Indiana University for his suggestions on the presentation of this work.

References

1. Kellis, M., Patterson, N., Endrizzi, M., Birren, B., Lander, E.: Sequencing and comparison of yeast species to identify genes and regulatory elements. Nature **423** (2003) 241–254
2. Pevzner, P., Tesler, G.: Human and mouse genomic sequences reveal extensive breakpoint reuse in mammalian evolution. Proc. Natl. Acad. Sci. U.S.A. **100** (2003) 7672–7677
3. Schwartz, S., Kent, W.J., Smit, A., Zhang, Z., Baertsch, R., Hardison, R.C., Haussler, D., Miller, W.: Human-mouse alignments with BLASTZ. Genome Res. **13** (2003) 103–107
4. Needleman, S.B., Wunsch, C.D.: A general method applicable to the search for similarities in the amino acid sequence of two proteins. J. Mol. Biol. **48** (1970) 443–453
5. Smith, T.F., Waterman, M.S.: Identification of common molecular sequences. J. Mol. Biol. **147** (1981) 195–197
6. Lipman, D.J., Altschul, S.F., Kececioglu, J.D.: A tool for multiple sequence alignment. Proc. Natl. Acad. Sci. U.S.A. **86** (1989) 4412–4415

7. Thompson, J., Higgins, D., Gibson, T.: CLUSTAL W: improving the sensitivity of progressive multiple sequence alignment through sequence weighting, position-specific gap penalties and weight matrix choice. Nucleic Acids Res. **22** (1994) 4673–4680

8. Corpet, F.: Multiple sequence alignment with hierarchical clustering. Nucleic Acids Res. **16** (1988) 10881–10890

9. Gotoh, O.: Significant improvement in accuracy of multiple protein sequence alignments by iterative refinement as assessed by reference to structural alignments. J. Mol. Biol. **264** (1996) 823–838

10. Notredame, C., Higgins, D.: SAGA: sequence alignment by genetic algorithm. Nucleic Acids Res. **24** (1996) 1515–1524

11. Kim, J., Pramanik, S., Chung, M.: Multiple sequence alignment using simulated annealing. Comput. Appl. Biosci. **10** (1994) 419–426

12. Höhl, M., Kurtz, S., Ohlebusch, E.: Efficient multiple genome alignment. Bioinformatics **18** (2002) S312–S320

13. Morgenstern, B., Frech, K., Dress, A., Werner, T.: DIALIGN: Finding local similarities by multiple sequence alignment. Bioinformatics **14** (1998) 290–294

14. Brudno, M., Do, C.B., Cooper, G.M., Kim, M.F., Davydov, E., NISC Comparative Sequencing Program, Green, E.D., Sidow, A., Batzoglou, S.: LAGAN and Multi-LAGAN: Efficient tools for large-scale multiple alignment of genomic DNA. Genome Res. **13** (2003) 721–731

15. Bray, N., Pachter, L.: MAVID: Constrained ancestral alignment of multiple sequences. Genome Res. **14** (2004) 693–699

16. Blanchette, M., Kent, W.J., Riemer, C., Elnitski, L., Smit, A.F., Roskin, K.M., Baertsch, R., Rosenbloom, K., Clawson, H., Green, E.D., Haussler, D., Miller, W.: Aligning multiple genomic sequences with the threaded blockset aligner. Genome Res. **14** (2004) 708–715

17. Schwartz, S., Elnitski, L., Li, M., Weirauch, M., Riemer, C., Smit, A., Program, N.C.S., Green, E.D., Hardison, R.C., Miller, W.: MultiPipMaker and supporting tools: alignments and analysis of multiple genomic DNA sequences. Nucleic Acids Res. **31** (2003) 3518–3524

18. Kim, S.: Graph theoretic sequence clustering algorithms and their applications to genome comparison. In Wu, C.H., Wang, P., Wang, J.T.L., eds.: Computational Biology and Genome Informatics. World Scientific (2003)

19. Kim, S., Gopu, A.: Cluster utility: A new metric to guide sequence clustering. Technical report, School of Informatics, Indiana University (2004)

20. Miller, W.: Comparison of genomic DNA sequences: Solved and unsolved problems. Bioinformatics **17** (2001) 391–397

On the Structure of Reconciliations*

Paweł Górecki[1] and Jerzy Tiuryn[1]

Warsaw University, Institute of Informatics, Banacha 2,
00-097 Warsaw, Poland
{gorecki,tiuryn}@mimuw.edu.pl

Abstract. In this paper we present an extended model related to reconciliation concepts. It is based on gene duplications, gene losses and speciation events. We define an evolutionary scenario (called a DLS-tree) which informally can represent an evolution of genes in species. We are interested in all scenarios - not only parsimonious ones. We propose a system of rules for transforming the scenarios. We prove that the system is confluent, sound and strongly normalizing. We show that a scenario in normal form (i.e. non-reducible) is unique and minimal in the sense of the cost computed as the total number of gene duplications and losses. Moreover, we present a classification of the scenarios and analyze their hierarchy. Finally, we prove that the tree in normal form could be easily transformed into the reconciled tree [12] in duplication-loss model. This solves some open problems stated in [13].

Keywords: molecular evolution, phylogenetic tree, reconciliation, gene duplication, gene loss, computational biology.

1 Introduction

The relationships between species cannot always be inferred from a single gene family. Gene duplication, gene loss, gene convergence, horizontal gene transfer and errors in sequencing could cause unexpected dissimilarities between gene family trees. Those incosistencies lead to the two important problems: reconstruction of the species tree from a family of possibly different gene trees and reconciling a given gene tree with a given species tree.

The problems have been studied in the seventies of 20th century by Goodman [6] and then in the nineties by Page, Guigó, Muchnik, Smith and others [5, 9, 11, 12, 14, 15]. The concepts of *mapping* and *reconciling trees* were introduced. They inspired research on *duplication-loss models* (we call them DL-models) and their extensions, for instance models with a horizontal gene transfer [3, 7, 10]. All DL-models are believed to be biologically meaningful [11].

Almost all approaches to reconstruction of evolution history are *parsimonious* i.e. it is assumed that the solution with the minimal cost is the most likely one. There are several possible cost functions: size of the reconstructed tree, or the number of specified evolutionary events, e.g. gene duplications, or the total number of gene duplications and gene losses. The latter measure, called *mutation cost*, was particularly popular among

---

* Financial support is provided by KBN Grant 4 T11F 020 25

J. Lagergren (Ed.): RECOMB 2004 Ws on Comparative Genomics, LNBI 3388, pp. 42–54, 2005.
© Springer-Verlag Berlin Heidelberg 2005

researchers [8, 12]. One of the crucial terms in the model is that of *reconciled tree* which represents the common evolutionary history of genes and species. In [2] authors present several definitions of a reconciled tree which were used recently in the literature. They proved that the definitions are equivalent and that the tree is minimal with respect to the size. Paper [2] still left open questions: (see [13]) *is the reconciled tree minimal with respect to the mutation cost* or *is it minimal with respect to the total number of gene duplications (duplication cost)*. Also the question of uniqueness of such a tree was left open. In the present paper we answer all these questions. We build a formal framework of evolutionary scenarios which represents a common history of genes and species under the assumption that only gene duplications, losses and speciations may occur. These scenarios are called here DLS-trees[1]. Given a DLS-tree T we show how to retrieve from T a gene tree *gene*(T), as well a species tree *spec*(T). A DLS-tree is similar to a concept of a reconciliation (see [1]) for a given species tree and a gene tree. We introduce a system of rules for transforming DLS-trees. This is a certain kind of a term rewrite system. It has pleasing mathematical properties: soundness[2], confluence and strong normalization. We prove that, a DLS-tree in normal form has minimal size, minimal mutation cost, and minimal duplication cost. It follows from our theory that for every DLS-tree T in normal form, if D_T is the set of all DSL-trees which have the normal form T, then T is the unique tree in D_T among all trees in D_T having the same mutation cost. We show an example that the uniqueness property fails when mutation cost is replaced by duplication cost (see Fig. 8). We show a one-to-one correspondence between the reconciled trees and the DLS-trees in normal form. Thus the theory build in this paper is immediately applicable to reconciled trees. We obtain a formula for computing the total number of duplications and losses in a reconciled tree, as a function of G and S. A formal analysis of these formulas in the context of reconciled trees can be found in [4, 16].

The paper is organized as follows first we define basic terms and DLS-trees. Then we show how to extract a gene and a species tree from a DLS-tree. In sections 4 and 5 we present the system of rules and prove soundness, completeness and confluence. In 6 we present an example of a hierarchy of DLS-trees together with all their reductions (Fig. 8). In section 7 we present formulas for computing the tree in normal form (for a given species tree and a given gene tree) and the number of duplications and losses. Finally, we show a one-to-one correspondence between the reconciled trees and the DLS-trees in normal form.

2   Gene and Species Trees

Let $\mathcal{I}$ be a finite set, called a set of species. A *gene tree* is a rooted binary directed tree whose leaves are labelled by the elements from $\mathcal{I}$. The labelling need not be one-to-one. A *species tree* is a gene tree[3] whose leaves are uniquely labelled.

Let T be a gene tree. For a node v of T by $T(v)$ we denote the subtree of T rooted in v. For each node v of a gene tree T we define a multiset $\mathfrak{m}_v^T = \{x_1^{i_1}, x_2^{i_2}, ..., x_k^{i_k}\}$, where

---

[1] DLS stands for **D**uplication, **L**oss and **S**peciation

[2] If T reduces to T' then *gene*$(T) =$ *gene*(T') and *spec*$(T) =$ *spec*(T')

[3] Although it seems to be strange from biological point of view this definition is mathematically correct. Here a species tree is a special case of a gene tree

for $j \in \{1, ..., k\}$ the upper index i_j of x_j is the number of leaves labelled x_j in $T(v)$ [4]. For the tree $\mathcal{G}$ presented in Fig. 2 we have $\mathfrak{m}_r = \{h^2, c^1, d^1\}$, where r is the root of $\mathcal{G}$. Analogously we define *a cluster* for v as a set $m_v^T = \{x_1, ..., x_m\}$. Let $\mathfrak{M}^T$ denote the multiset $\{\mathfrak{m}_v^T | v \in V\}$. In order to make the notation more readable m_v^T will be denoted by $x_1...x_m$. Note that if T is a species tree then $\mathfrak{m}_v^T = m_v^T$. We denote by $root(T)$ the root of T and by $L(T)$ the set of all labels of leaves in T.

A multiset M is said to *determine a species tree* if $\mathfrak{M}^S = M$, for some species tree S. It can be proved that M determines a species tree if only if

(M1) if M is nonempty, then $\bigcup M \in M$,
(M2) for each $a \in \bigcup M$, $\{a\} \in M$,
(M3) for each $A \in M$ such that A is not a singleton, $\{X | X \in M \text{ and } X \subsetneq A\}$ contains two maximal (in the sense of inclusion) sets.

It can be shown that if T and S are species trees and $\mathfrak{M}^T = \mathfrak{M}^S$ then $T = S$. The above property does not hold in general for gene trees.

We use the standard *nested-parenthesis notation* for trees i.e.:

- The empty tree will be denoted by $\emptyset$,
- The label a denotes a tree with one node labelled by a,
- If T_p and T_q are two nonempty trees with roots p and q, respectively, then (T_p, T_q) is a tree whose root has two children: p and q. The trees T_p and T_q are rooted in (T_p, T_q) at the nodes p and q, respectively.

3   DLS Trees

Now we define a crucial notion of a DLS-tree. Such a tree could be interpreted as "an evolutionary scenario representing history of genes in the context of species evolution".

First we start with some biological motivations. Fig. 1 presents all aspects of the common evolution of genes and species under assumption that only gene duplications, gene losses or speciations can occur.

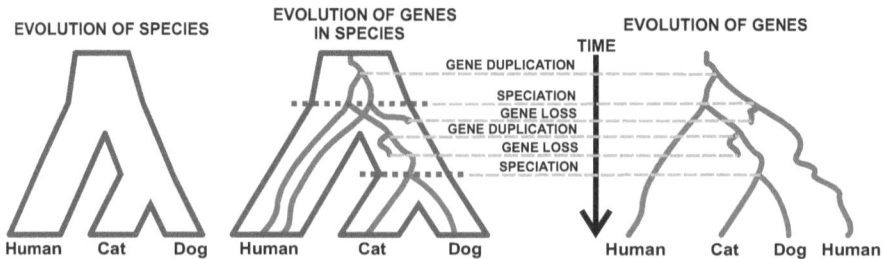

Fig. 1. Evolution of species and genes

The left tree presents an evolutionary species tree and its interpretation is clear. The rightmost tree presents an evolution of a family of genes which are related to the

---

[4] We assume that $i_j > 0$

three species. Note, that we have two genes labelled by the species *Human*. It means that both genes are currently present in human. This situation is a consequence of the first *gene duplication* which happened early in the evolution when only one species existed i.e. the common ancestor of *human, cat* and *dog*. We see also that some of the gene lineages are lost. Here we have two gene losses. Current methods of gene tree reconstruction (from gene sequences) cannot detect this kind of losses which are shown in this Figure. Although if we know the species tree and the gene tree we can find evolutionary scenarios which explain the differences between the tree in terms of gene duplications and losses. One of them is shown in the middle tree (Fig. 1). We see the embedding of the gene tree (right) into the species tree (left). It is clear that this kind of embedding is biologically correct. Note, that the internal nodes of the gene tree are related either to speciations or to gene duplications.

Our goal is to present a mathematical model of the evolutionary scenario. Let us adopt the following symbols □ (duplication), ○ (loss), ▬ (speciation) and ● (gene).

A *DLS-tree* is either an empty tree, or a binary rooted tree $T = (V, E)$ such that the elements of V are labelled by nonempty subsets of $\mathcal{I}$. For $v \in V$ let Λ_v denote the label of v. V is divided into four disjoint sets $V_\bullet$, $V_\bigcirc$, $V_\square$ and $V_\blacksquare$ such that[5]

(D1) if $v \in V_\bullet$ then v is a leaf in T labelled by a species a (v is called a gene node),
(D2) if $v \in V_\bigcirc$ then v is a leaf in T (v is called a loss node),
(D3) if $v \in V_\square$ then v has two children a and b such that $\Lambda_a = \Lambda_b = \Lambda_v$ (v is called a duplication node),
(D4) if $v \in V_\blacksquare$ then v has two children a and b such that $\Lambda_a \cup \Lambda_b = \Lambda_v$ and $\Lambda_a \cap \Lambda_b = \emptyset$ (v is called a speciation node),
(D5) for all $v, w \in V$ such that $\Lambda_v \cap \Lambda_w \neq \emptyset$ we have either $\Lambda_v \subseteq \Lambda_w$ or $\Lambda_v \supseteq \Lambda_w$.

By $\mathfrak{Labels}^T$ we denote the set of all labels in T.

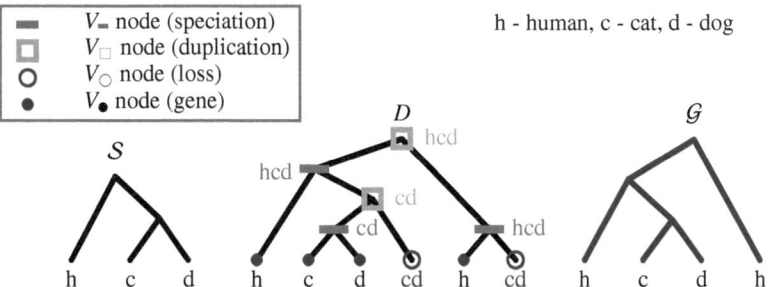

Fig. 2. Species (S), DLS (D) and gene tree (G) for the example shown in Fig. 1

Compare Fig. 2 which presents the trees in our model with Fig. 1. With a DLS-tree T we associate *a cost* which is the total number of gene duplications and losses in T. This cost is known in the literature as *a mutation cost* [8].

[5] Sometimes we use an upper index to distinguish objects from different trees

Sometimes we will use a linear (term-like) representation of DLS-trees (we omit the formal definition). For example the tree D in Fig. 2 can be described as $((h, ((c, d)_\blacksquare, cd_\bigcirc)_\square)_\blacksquare, (h, cd_\bigcirc)_\blacksquare)_\square$.

3.1   Extracting Gene and Species Trees from DLS Trees

We explain how to extract from a given DLS-tree a gene tree, and a species tree, relying on information contained in its labels.

We start with the gene tree. For a set of leaves L in T let T^L be the smallest subtree of T containing L as its set of leaves. The *homomorphic tree* $T|_L$ of T induced by L is the tree obtained from T^L by contracting all nodes of degree 2 except for its root (i.e. for each such a node x: create an edge connecting the parent of x with the only one child of x; remove x and all edges incident on it) [2,9]. Now we can use the homomorphic tree to get the gene tree from a DLS-tree. Let T be a DLS-tree. We set $gene(T)$ to be the gene tree defined by $T|_{V_\bullet}$. The labels of leaves in $gene(T)$ are inherited from T. One can easily check that for the trees from Fig. 2 we have $gene(D) = \mathcal{G}$.

Now we present the extraction of the species tree. The natural question is whether the set $\mathfrak{Labels}$ determines a species tree. Fig. 3 presents a DLS-tree which does not satisfy this property. We see that the tree contains an incomplete information on a species relationship due to the loss nodes. To solve the problem we have to identify species for which the reconstruction (from labels) will give a species tree. We call a species s *lost in* T if s occurs only in loss nodes of T. Formally the set of lost species can be defined by $\mathfrak{lost}^T = L(T) \setminus \bigcup \{ \Lambda_v^T | v \in V_I^T \}$ (see Fig. 3 for example). It can be proved that if we remove lost species from all labels of a DLS-tree then we will be able to reconstruct the species tree. Formally we claim that $\{ \Lambda_v \setminus \mathfrak{lost}^T | v \in T \} \setminus \{ \emptyset \}$ determines a species tree. We denote the tree by $spec(T)$. Let us notice that for a DLS-tree T we have $L(gene(T)) = L(spec(T))$.

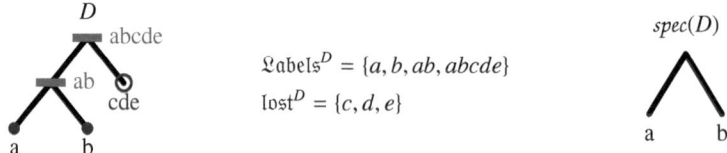

$$\mathfrak{Labels}^D = \{ a, b, ab, abcde \}$$
$$\mathfrak{lost}^D = \{ c, d, e \}$$

Fig. 3. Incomplete DLS-tree D

The species tree for the incomplete DLS-tree D is presented in Fig. 3.

We call a DLS-tree *complete* if the tree has no lost species. Fig. 4 presents a complete DLS-tree.

4   DLS Rules

We define DLS rules (we call them rules). They will be used to transform DLS-trees. Each rule is defined by $\frac{P}{Q}$ where P (premise) and Q (conclusion) are DLS-trees. By a redex of a rule R we mean a node v in a tree to which the premise of R is applicable. A DLS-tree T can be transformed into T' by a rule R in node v iff

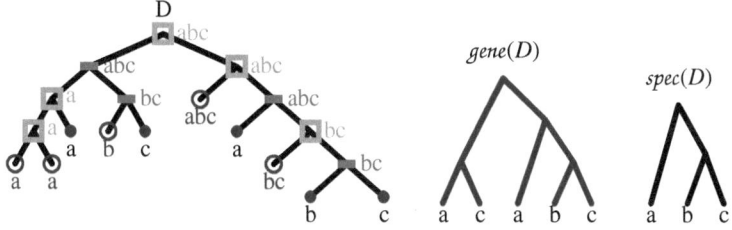

Fig. 4. A complete DLS-tree D and extracted gene and species trees. The cost equals 10

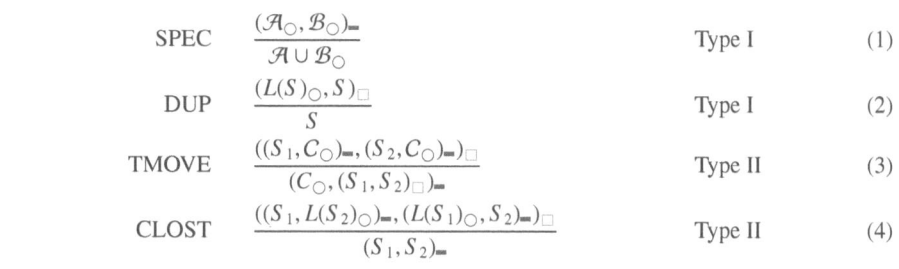

Fig. 5. DLS rules

- P equals $T(v)$,
- T' is constructed from T by replacing this subtree by the tree Q.

Let $R(T, v)$ denote the result of reduction.

We write $T \to T'$ if T' is constructed from T by an application of one rule. We write $T \twoheadrightarrow T'$ if T' is constructed from T by applying zero or more rules. We use $\to^{-1}$ to denote backward transformation i.e. $T' \to^{-1} T$ iff $T \to T'$. The rules are presented in Fig. 5 and their graphical interpretation in Fig. 6. It should be clear that an application of any rule to a DLS-tree yields a DLS-tree again and the reduction decreases the cost.

The following Lemma states the soundness of the system:

Proposition 1. *(Soundness) If $T \to T'$ then $gene(T) = gene(T')$ and $spec(T) = spec(T')$.*

5   Properties of the System

In this section we present some important properties of DLS-trees. A DLS-tree containing no type I redexes is called *a semi-normal* tree.

A semi-normal tree T is called *fat* if the following conditions are satisfied

- every duplication node has label $L(T)$.
- each speciation node has exactly one lost child.

Fig. 7 presents an example of a fat DLS-tree.

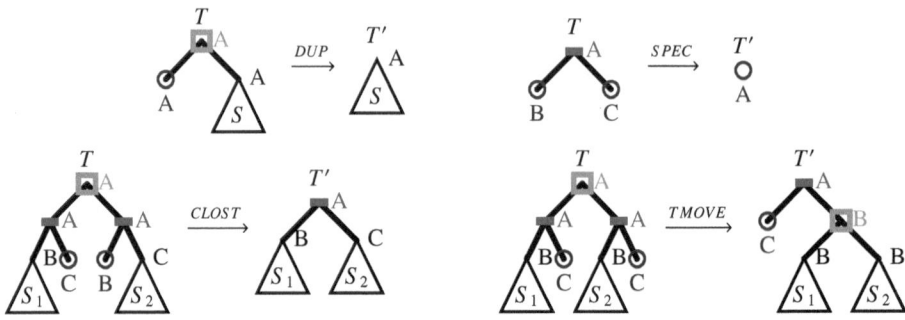

Fig. 6. DLS rules

It can be proved that for a fat tree each child of a duplication node is either a duplication node, or is the root of a tree of the form:

$$(B_{1\bigcirc}, (B_{2\bigcirc}, ...(B_{k\bigcirc}, a)_{\blacksquare}...)_{\blacksquare})_{\blacksquare},\tag{5}$$

where $a \in I$ and $k \geq 0$. We call the tree (5) a *chain tree*. Its *target* is a.

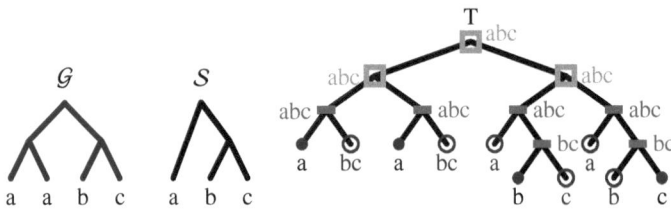

Fig. 7. A fat DLS-tree T, its gene tree G and species tree S

Extracting gene and species trees from fat trees is quite natural (see Fig. 7):

Proposition 2. *Let us assume that T is fat. Then*

(1) gene(T) is constructed from T by replacing each chain tree in T by a single node labelled by its target,

(2) if there are no lost species in T then spec(T) is determined by $\bigcup_{C \in C(T)} \mathfrak{M}^C$ where C(T) denotes the set of all chain trees in T.

Also converse holds:

Proposition 3. *Given a gene tree G and a species tree S such that $L(G) = L(S)$. There exists a unique fat tree T such that gene(T) = G and spec(T) = S.*

Proof. For each label $a \in L(G)$ we define the chain tree S_a with a target a. Let $p_0...p_k$ be the (unique) path in S such that $p_0 = root(S)$ and p_n has label a. Let S_a be the chain tree (5) such that $B_i = m^S_{p_{i-1}} \setminus m^S_{p_i}$ for $i = 1, ..., k$. Note, that $\Lambda^{S_a}_{root(S_a)} = L(G)$.

Now we show the construction of the fat tree T. We transform each internal node of the gene tree into a duplication node labelled by $L(G)$. Each leaf labelled by a in the tree is replaced by S_a. We claim, that T is correctly defined. Details are omitted. $\square$

We define $\sim$ to be the least equivalence relation in the set of DLS-trees which contains relation $\rightarrow$. Thus, if $T \sim T'$ then T can be transformed into T' by applying DLS rules zero or more times in any direction.

Proposition 4. *Every DLS-tree is equivalent to a unique fat tree.*

Proof. Apply the following procedure:
(1) First we eliminate iteratively all redexes of the rules DUP and SPEC.
(2) Eliminate all redexes of TMOVE in a reverse direction.
(3) Eliminate all redexes of CLOST in a reverse direction.
      We claim that after finishing this procedure we get a unique fat tree. $\square$

Observe that we can increase the cost of a fat tree by applying SPEC in direction $\rightarrow^{-1}$; in this way we increase each $B_\bigcirc$ by at most $|B| - 1$, or by applying DUP in direction $\rightarrow^{-1}$; this can be done an unbounded number of times, increasing the number of the duplication nodes an introducing spurious loss nodes $\mathcal{I}_\bigcirc$.

Note, that applying transformations (2) and (3) (see proof of Prop. 4) we get a tree with a larger size. Thus we conclude that a fat tree is the heaviest (in the sense of size) among all equivalent semi-normal trees.

We can also prove the completeness of the system. Recall that a complete DLS-tree is a tree without lost species:

Proposition 5. (Completeness) *Let T_1 and T_2 be complete DLS-trees such that $gene(T_1) = gene(T_2)$ and $spec(T_1) = spec(T_2)$. Then $T_1 \sim T_2$.*

Proof. By Proposition 4 there exist fat trees T_1' and T_2' such that $T_1 \sim T_1'$ and $T_2 \sim T_2'$. By Prop. 1 for $i = 1, 2$ we have $gene(T_i) = gene(T_i')$ and $spec(T_i) = spec(T_i')$. By the assumption and Proposition 3 we get $T_1' = T_2'$ (from the uniqueness of the fat tree). $\square$

The following Proposition states that the system is weakly confluent. We omit the technical proof which requires analysis of all possible cases.

Proposition 6. *Let T be a DLS-tree. Then for each T_1 and T_2 such that $T \rightarrow T_1$ and $T \rightarrow T_2$ there exists T_3 such that $T_1 \twoheadrightarrow T_3$ and $T_2 \twoheadrightarrow T_3$.*

Theorem presented below states that our system is confluent.

Theorem 1. (Confluence) *Take a DLS-tree T. There exists a unique DLS-tree T^* (in normal form) such that every sequence of reductions in direction $\rightarrow$ yields T^*.*

Proof. The termination follows from the fact that every application of rules reduces the cost.

Let us assume that T_1^* and T_2^* are in normal form such that $T \twoheadrightarrow T_1^*$ and $T \twoheadrightarrow T_2^*$. Obviously T_1^* and T_2^* are semi-normal. From 1 and the proof of Prop. 4 we conclude that there exists a unique fat tree T^f such that $T^f \twoheadrightarrow T_1^*$ and $T^f \twoheadrightarrow T_2^*$. Thus by Prop. 6 there exists T^* such that $T_1^* \twoheadrightarrow T^*$ and $T_2^* \twoheadrightarrow T^*$. But both trees are in normal form thus $T_1^* = T_2^* = T^*$. $\square$

Theorem 2. *For DLS-trees T_1 and T_2 we have $T_1 \sim T_2$ iff $T_1^* = T_2^*$.*

Proof. (=>). Obviously $T_1^* \sim T_2^*$. From the proof of Prop. 4 and Cor. 1 we conclude that there exists a unique fat tree T^f such that $T^f \twoheadrightarrow T_1^*$ and $T^f \twoheadrightarrow T_2^*$. By Thm. 1 there exists a unique tree T^* in normal form such that $T^f \twoheadrightarrow T^*$. Finally, by Prop. 6 we get $T_1^* = T_2^* = T^*$.
(<=). We have $T_1 \twoheadrightarrow T_1^*$ and $T_2 \twoheadrightarrow T_2^*$. Thus $T_1 \sim T_2$.   □

Corollary 1. *For a DLS-tree T, T^* is the unique tree with minimal cost in the set of all trees which are equivalent to T.*

6   Semi-normal Trees

As we noticed the semi-normal trees are important representants of each class of equivalent DLS-trees. In this section we consider a hierarchy of equivalent semi-normal trees and summarize its properties.

By the proof of Prop. 4 and further discussion we can transform each semi-normal tree into the unique fat tree in two steps. In the first step we apply all possible CLOST rules in the reverse direction. Then we apply TMOVE rules in the reverse direction.

Analogously, we can transform each semi-normal tree into the unique tree in a normal form. First we apply TMOVE rules, then CLOST in the direction $\rightarrow$.

An example of a hierarchy of semi-normal trees with all possible reductions is presented in Fig. 8. In our example we have all possible 10 semi-normal DLS-trees. T_f is the fat tree and T^* is the tree in normal form. The labels of the internal nodes are not shown. They can be easily reconstructed from the labels of the leaves. The nodes marked by v_i are the redexes of the rules. For instance CLOST(v_4) above the arrow from T_2 to T_5 denotes the equation $T_5 = \text{CLOST}(T_2, v_4)$.

Note that, if we consider only the duplication cost we loose the uniqueness of the minimal tree In Fig. 8 the trees T^* and T_9 have the same duplication cost i.e. 2.

7   Tree in Normal Form

In this section we present the construction of a DLS-tree in normal form $\rho(\mathcal{G}, \mathcal{S})$ for a given species tree $\mathcal{S}$ and a gene tree $\mathcal{G}$, subject to the condition $\emptyset \neq L(\mathcal{G}) \subseteq L(\mathcal{S})$.

Let $\rightsquigarrow$ denote a path existence relation in $\mathcal{S}$ i.e. $a \rightsquigarrow b$ iff there exists a path from a to b in $\mathcal{S}$. Let $\rightarrow$ denote a child relation i.e. $a \rightarrow b$ iff b is a child of a. Reversed arrows will be used to denote the reversed relations.

Let $g \in \mathcal{G}$. By $M(g)$ we denote the node $s \in \mathcal{S}$ such that

$$m_s^S = \bigcap \{m_w^S \mid m_g^G \subseteq m_w^S\}.$$

The obtained function $M : \mathcal{G} \rightarrow \mathcal{S}$ is called in the literature [6, 12] *a least common ancestor mapping* or shorter *lca-mapping* (see Fig. 9).

The definition of $\rho(\mathcal{G}, \mathcal{S})$ is by structural induction on the size of $\mathcal{G}$ and $\mathcal{S}$. Let $s = root(\mathcal{S})$ and $g = root(\mathcal{G})$. If $\mathcal{S}$ and $\mathcal{G}$ are leaves then $\rho = a$, where a is the label of g. Otherwise, let p and q be the children of g then

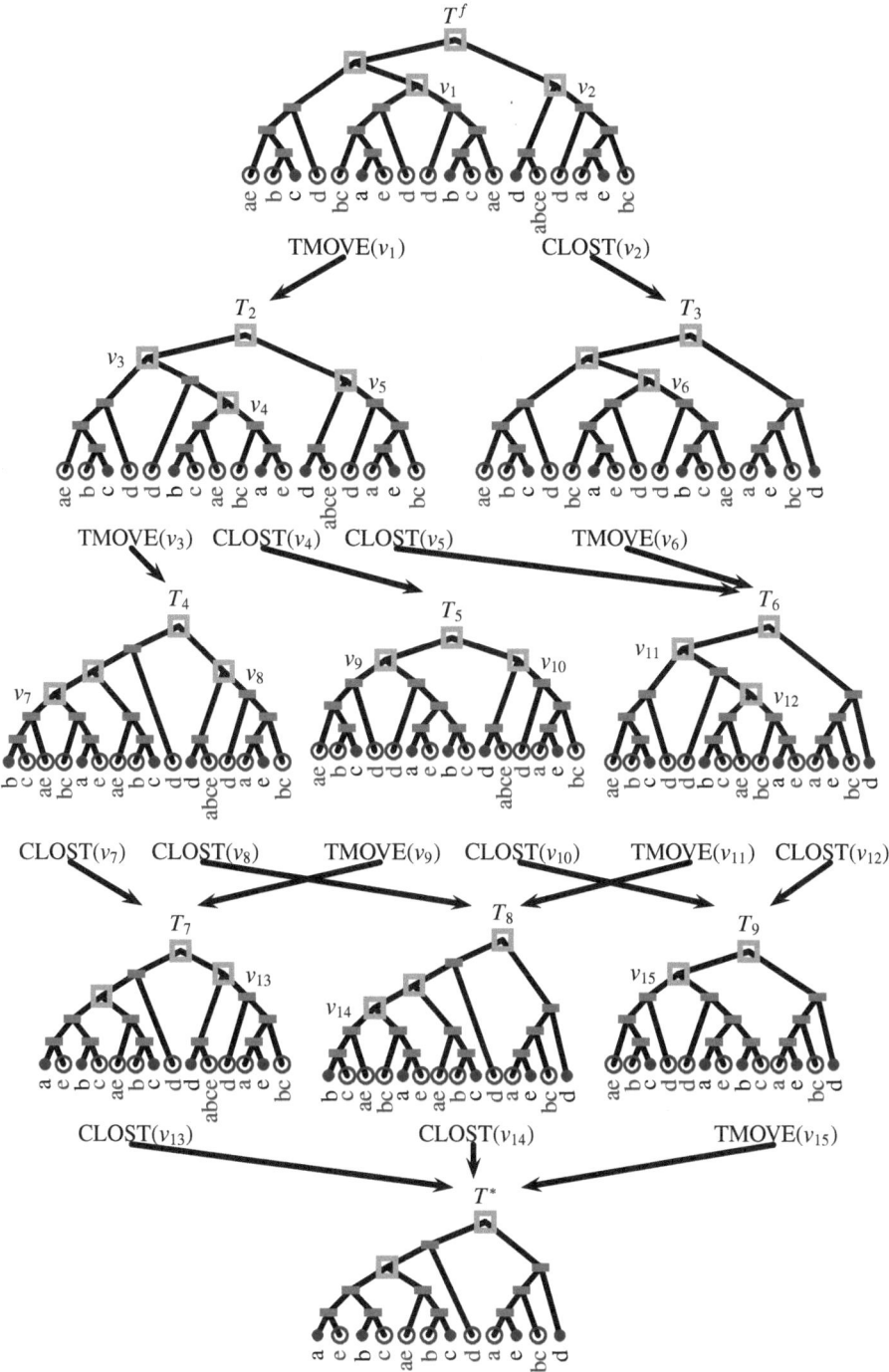

Fig. 8. Example hierarchy of semi-normal trees with all possible reductions

$$\rho(G, S) = \begin{cases} (\rho(G(p), S), \rho(G(q), S))_{\square} & \text{if } M(g) = s = M(q), & (DUP) \\ (\rho(G(p), S(a), \rho(G(q), S(b))))_{\blacksquare} & \text{if } M(p) \leftsquigarrow a \leftarrow s = M(g) \\ & \text{and } s \to b \rightsquigarrow M(q), & (SPEC) \\ (\rho(G, S(a)), m^S_{b\bigcirc})_{\blacksquare} & \text{if } M(g) \leftsquigarrow a \leftarrow s \to b \neq a. (LOSS) \end{cases} \quad (6)$$

We claim that Equation (6) correctly defines a DLS-tree. One of the most important properties of ρ is stated below:

Lemma 1. *Let G be a gene tree and S be a species tree such that $\emptyset \neq L(G) \subseteq L(S)$. Then $\rho(G, S)$ is in normal form.*

Again, we omit the details of the proof. Now, we conclude that if T is a complete DLS-tree then $T^* = \rho(gene(T), spec(T))$ [6].

Having formula (6) we can compute the number of evolutionary events in a tree in normal form. We can prove the following Lemmas:

Lemma 2. *Let G be a gene tree and S be a species tree such that $\emptyset \neq L(G) \subseteq L(S)$. Then the number of duplications in $\rho(G, S)$ equals*

$$\mathbf{dup}(G, S) = |\{g \mid M(g) = M(p) \text{ where } p \text{ is a child of } g \text{ in } G\}|. \quad (7)$$

Lemma 3. *Let G be a gene tree and S be a species tree such that $\emptyset \neq L(G) \subseteq L(S)$. For each node g, we define a non-negative integer $\mathbf{loss}_g$ as follows: we set $\mathbf{loss}_g = 0$ if g is leaf in G, if g is an internal node in G then let p and q denote the two children of g. We define*

$$\mathbf{loss}_g = \begin{cases} d(g, p) + 1 & \text{if } M(p) \neq M(g) = M(q), \\ d(g, p) + d(g, q) & \text{otherwise.} \end{cases} \quad (8)$$

where $d(g, g') = |\{s \mid m^S_{M(g')} \subsetneq m^S_s \subsetneq m^S_{M(g)}\}|$.
Then the number of gene losses in $\rho(G, S)$ is given by $\sum_{g \in G} \mathbf{loss}_g$.

Now we present a definition of the reconciled tree taken from [2]. We know (see [2]) that this definition is equivalent to the definition given by Page [12]. Let $s = root(S)$ and $g = root(G)$. The reconciled tree $R(G, S)$ of G with respect to S is the tree G if G and S are leaves. Otherwise, let p and q are the children of g, then

- $(R(G(p), S), R(G(q), S))$ if $M(g) = s = M(q)$,
- $(R(G(p), S(a)), R(G(q), S(b)))$ if $M(p) \leftsquigarrow a \leftarrow s = M(g) \to b \rightsquigarrow M(q)$,
- $(R(G, S(a)), S(b))$ if $M(g) \leftsquigarrow a \leftarrow s \to b \neq a$

Theorem 3. *Let G be a gene tree and S a species tree such that $\emptyset \neq L(G) \subseteq L(S)$. We define a transformation θ which takes one argument: a DLS-tree and returns a reconciled tree. Let*

- $\theta(A_\bigcirc) = S(v)$ *where v in S such that $m^S_v = A$,*
- $\theta(a) = a$ *where $a \in I$,*
- $\theta((T_1, T_2)_*) = (T_1, T_2)$ *where $* \in \{\blacksquare, \square\}$.*

Then $\theta(\rho(G, S)) = R(G, S)$.

---

[6] This result can be extended to any DLS-tree. Here we omit the discussion

The easy proof of Theorem 3 follows immediately from the definition of the rec-onciled tree and the definition of ρ. We prove a one-to-one correspondence between DLS-trees in normal form and reconciled trees. Moreover, we claim that the formula for computing the mutation cost is the same for the reconciled tree [8] and the tree in normal form.

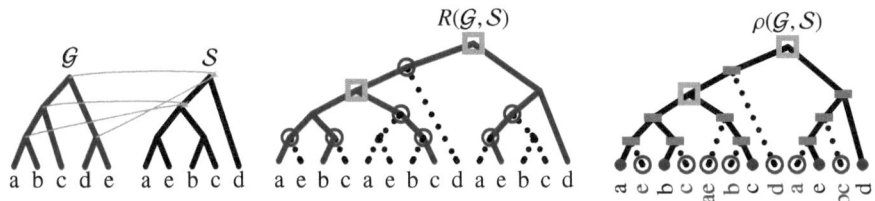

Fig. 9. Example of mapping M (only for internal nodes of $\mathcal{G}$), reconciled tree $R(\mathcal{G}, \mathcal{S})$ and DLS-tree $\rho(\mathcal{G}, \mathcal{S})$

Fig. 9 presents an example of a lca mapping M for the internal nodes of $\mathcal{G}$. It also presents a reconciled tree $R(\mathcal{G}, \mathcal{S})$ and a DLS-tree $\rho(\mathcal{G}, \mathcal{S})$. Note that $\rho(\mathcal{G}, \mathcal{S})$ equals the tree T^* from Fig. 8. It is easy to notice that $\theta(\rho(\mathcal{G}, \mathcal{S})) = R(\mathcal{G}, \mathcal{S})$. For a more readable presentation all the lost gene lineages are shown with dotted lines. The solid lines in $R(\mathcal{G}, \mathcal{S})$ and $\rho(\mathcal{G}, \mathcal{S})$ represent embedded gene trees.

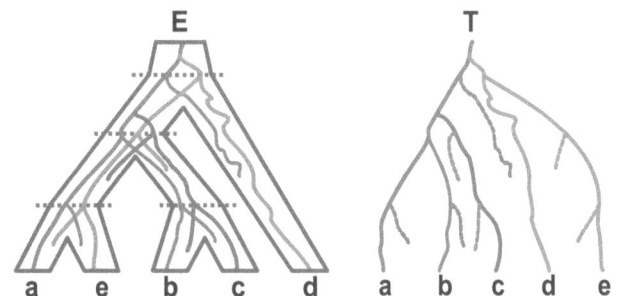

Fig. 10. The evolution of species and genes (cont. example from Fig. 9)

Fig. 10 is a continuation of the example presented in Fig. 9. Tree E is the final embedding of $\mathcal{G}$ into $\mathcal{S}$. Tree T presents an extraction of gene lineages from E. Note that the tree T is equal topologically to the DLS-tree $\rho(\mathcal{G}, \mathcal{S})$.

References

1. L. Arvestad, A.-C. Berglund, J. Lagergren, and B. Sennblad. Gene tree reconstruction and orthology analysis based on an integrated model for duplications and sequence evolution. In *RECOMB 2004*, 2004.

2. P. Bonizzoni, G.D. Vedova, and R. Dondi. Reconciling gene trees to a species tree. *Algorithms and Complexity, Proceedings of the 5th Italian Conference (CIAC 2003)*, Vol. 2653:120–131, 2003.
3. M. A. Charleston. Jungles: A new solution to the host/parasite phylogeny reconciliation problem. *Mathematical Biosciences*, 149:191–223, 1998.
4. Eulenstein and Vingron. On the equivalence of two tree mapping measures. *DAMATH: Discrete Applied Mathematics and Combinatorial Operations Research and Computer Science*, 88, 1998.
5. O. Eulenstein, B. Mirkin, and M. Vingron. Duplication-based measures of difference between gene and species trees. *Journal of Computational Biology*, 1:135–148, 1998.
6. M. Goodman, J. Czelusniak, G. W. Moore, A. E. Romero-Harrera, and G. Matsuda. Fitting the gene lineage into its species lineage. A parsimony strategy illustrated by cladograms constructed from globin sequences. *Syst. Zool.*, 28:132–163, 1979.
7. M. T. Hallett and J. Lagergren. New algorithms for the duplication-loss model. In *Proceedings of RECOMB 2000*, pages 138–146, Tokyo, 2000. ACM Press.
8. Bin Ma, Ming Li, and Louixin Zhang. On Reconstructing Species Trees from Gene Trees in Term of Duplications and Losses. In *RECOMB 1998*, pages 182–191, 1998.
9. Bin Ma, Ming Li, and Louxin Zhang. From gene trees to species trees. *SIAM Journal of Comput.*, 30:792–752, 2000.
10. P. Górecki. Reconciliation problems for duplication, loss and horizontal gene transfer. In *RECOMB 2004*, San Diego, 2004.
11. B. Mirkin, I. Muchnik, and T. F. Smith. A biologically meaningful model for comparing molecular phylogenies. *J. of Comput. Biol.*, 2:492–507, August 9 1995.
12. R. D. M. Page. Maps between trees and cladistic analysis of historical associations among genes, organisms, and areas. *Systematic Biology*, 43:58–77, 1994.
13. R. D. M. Page and M. A. Charleston. Reconciled trees and incongruent gene and species trees. *Mathematical Hierarchies and Biology, DIMACS Series in Mathematics and Theoretical Computers Science*, 37(57–70), 1997.
14. R. D. M. Page. Component analysis: A valiant failure? *Cladistics*, 6:119–36, 1990.
15. I. Muchnik R. Guigo and T. Smith. Reconstruction of ancient molecular phylogeny. *Mol. Phy. and Evol.*, 6:189–213, August 9 1996.
16. L. Zhang. On a mirkin-muchnik-smith conjecture for comparing molecular phylog enies. *Journal of Computational Biology*, 4(2):177–188, 1997.

The Statistical Significance of Max-Gap Clusters

Rose Hoberman[1,*], David Sankoff[2], and Dannie Durand[3]

[1] Computer Science Department, Carnegie Mellon University, Pittsburgh, PA, USA
roseh@cs.cmu.edu
[2] Department of Mathematics and Statistics, University of Ottawa, Ontario, Canada
sankoff@uottawa.ca
[3] Departments of Biological Sciences and Computer Science,
Carnegie Mellon University, Pittsburgh, PA, USA
durand@cmu.edu

Abstract. Identifying *gene clusters*, genomic regions that share local similarities in gene organization, is a prerequisite for many different types of genomic analyses, including operon prediction, reconstruction of chromosomal rearrangements, and detection of whole-genome duplications. A number of formal definitions of gene clusters have been proposed, as well as methods for finding such clusters and/or statistical tests for determining their significance. Unfortunately, there is very little overlap between previously published rigorous analytical statistical tests and the definitions used in practice. In this paper, we consider the *max-gap* cluster: a contiguous region containing a maximal set of homologs, where the number of non-homologous genes between pairs of adjacent homologs is never greater than a predefined, fixed parameter, g. Although this is one of the models most widely used in practice, currently the statistical significance of max-gap clusters can only be evaluated using Monte Carlo simulations because no analytical statistical tests have been developed for it. We give exact expressions for the probability of observing such a cluster by chance, assuming a simple reference-region scenario and random gene order, as well as more efficient methods for approximating this probability. We use these methods to identify which regions of the parameter space yield clusters that are statistically significant. Finally, we discuss some of the challenges in extending this model to whole-genome comparison.

1 Introduction

Identification of conserved chromosomal segments is an essential first step for many different types of genomic analyses. Regions of similar gene content in related genomes can provide evidence for evolutionary relatedness or functional selection on gene order. For example, within a single genome the pattern of duplicated regions can provide evidence for large-scale or whole-genome duplication [2, 17, 18, 40, 53, 54, 66–68]. Conserved segments between different genomes, on the

---

* Contact author

J. Lagergren (Ed.): RECOMB 2004 Ws on Comparative Genomics, LNBI 3388, pp. 55–71, 2005.
© Springer-Verlag Berlin Heidelberg 2005

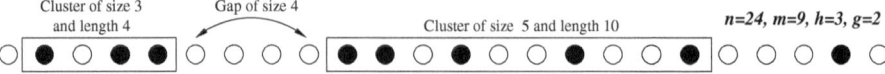

Fig. 1. A sample genome ($n = 24$), with $m = 9$ genes of interest shown in black. When the maximum gap allowed is $g = 2$ and the minimum cluster size is $h = 3$, then two clusters are found. The rightmost black gene is not part of any cluster.

other hand, have been used extensively to reconstruct the history of chromosomal rearrangements and infer an ancestral genetic map for a diverse group of species [8, 11, 16, 43, 41, 55, 47, 51], as well as to provide coarse-grain features for new phylogenetic approaches [6, 12, 26, 49, 50, 62]. In bacteria, conserved gene order and content have been used for prediction of operons [7, 20], horizontal transfers [36], and more generally to help understand the relationship between spatial organization and functional selection [31, 35, 45, 60, 61].

The common goal in all of these analyses is either to detect regions that share a common ancestor or where gene content is under functional selection. The signature of such conserved regions, which we call *gene clusters*, will be similar gene content, but we do not require gene content or order be strictly conserved as this would rule out many more distantly related regions.

It is not obvious how to choose a formal definition that best captures our intuitive notions about gene clusters. A number of definitions have been proposed, as well as algorithms for finding clusters which meet these definitions and statistical tests to evaluate their significance [3, 22, 23, 29]. The most stringent of these define conserved segments as two or more contiguous regions that contain the same genes in the same order [42, 44] and sometimes orientation [45, 60, 68]. However, such stringent definitions will invariably lead to the exclusion of many regions that did indeed descend from a single ancestral region but have since undergone small rearrangements. More flexible definitions allow for some amount of divergence and rearrangement.

Many of these more flexible definitions are based on a simple model in which a genome is represented as an ordered set of n genes: $G = (g_1, ..., g_n)$. Chromosome breaks are ignored and it is assumed that genes do not overlap. We start with a simple abstraction in which m genes ("the black genes") are pre-specified as interesting. These m genes may be of interest because their homologs are contiguous in another region or genome (the "reference region") or because they share some functional properties. We are interested in finding a large group of black genes that appear in close proximity. The *size* of the cluster is usually quantified as the total number of black genes in the cluster, where a *complete* cluster contains all m black genes and an *incomplete* cluster contains only a subset of the black genes. For example, a short genome with $n = 24$ genes is illustrated in Figure 1. The $m = 9$ black genes are shown grouped into two incomplete clusters, of size three and five respectively.

Although it is quite clear how to characterize cluster size, there is no agreed upon definition of "close proximity." Some definitions restrict the total *length* of the cluster [15] (the total number of genes from the first to the last black gene in the cluster). Others constrain the cluster density (the proportion of black genes in the cluster, or size / length). Others require only that clusters be compact [52], where compactness is determined by the distance, or *gap*, between adjacent black genes, that is, the number of white genes between them. For example, in Figure 1 the gap between the first and second black genes is one and the gap between the second and third black genes is zero. Of the definitions that constrain the gap sizes, some allow no gaps in a cluster [27, 28], others limit the sum of all gaps, while the majority constrain the size of the largest gap observed [5, 10, 40, 45, 56, 60, 65, 67].

In addition to cluster size and length, many cluster definitions constrain gene order, with some requiring a strictly conserved gene order, while others allow only a fixed number of order violations [25]. The majority ignore gene order altogether.

Although a number of formal definitions of gene clusters have been proposed, there is unfortunately very little overlap between cluster definitions used in analyses of genomic data and the definitions upon which rigorous analytical statistical tests are based. In this paper, we focus on a particular cluster definition that is widely used in genomic studies, including the identification of large-scale duplications in *Arabidopsis* [5] and the chordate lineage [40], the assignment of functions to uncharacterized genes in prokaryotes [45, 60], and the prediction of putative operons in newly sequenced bacterial genomes [10]. According to this definition, gene order is disregarded, and there is no limit on the total number of gaps as long as the maximum gap between adjacent black genes in the cluster is not too large. To distinguish these clusters from our informal notion of a cluster we call them *max-gap* clusters. A max-gap cluster is a maximal set of black genes where the gap between adjacent black genes is never larger than g. For example, when the maximum gap allowed is $g = 2$, three clusters can be found in the example genome in Figure 1. The first has size three and length four, the second has size five and length ten, and the third is a singleton.

The max-gap cluster definition has a number of desirable properties. It is flexible in that it does not require that every gene in the cluster have a homolog, yet it guarantees that the gap between adjacent homologs will not be too large. As a result, the density of a cluster is guaranteed to be no less than $1/(g + 1)$. This definition does not arbitrarily constrain the cluster length, but instead lets clusters grow to their "natural" size. Consequently, clusters will never overlap: unlike some other cluster definitions [15, 9], a gene can never be considered part of two distinct clusters that cannot be merged. On the other hand, two max-gap clusters containing the same number of homologs may have significantly different densities. For example, the length of a cluster of size m can range from m (density of one) to $g(m - 1) + m$ (a density close to $1/(g + 1)$). Finally, an algorithm has been developed for finding max-gap clusters efficiently [4]. However, most groups do not describe in detail the algorithm they use for finding max-gap clusters, so it is not clear whether they are using an efficient or even a correct algorithm.

Analytical statistical models in the literature are designed for other definitions of gene clusters [9, 14–16, 63, 66] and it is not obvious how to extend them to apply to this commonly used cluster model. Studies based on the max-gap cluster model usually use randomization to estimate the significance of clusters [5, 40, 45, 56, 65, 67]. However, this approach "is computationally expensive and does not permit very precise estimation of the probabilities of rare events" [9]. In addition, parameter values such as the maximum gap and minimum cluster size are generally selected in an ad-hoc manner. A formal, rigorous mathematical model of gene clusters will allow us to evaluate cluster significance more accurately and more quickly, and to choose parameter values in a principled manner.

Our goal in this paper is to try to close the gap between rigorous mathematical models and models used in the analysis of real genomes by developing formal statistical tests for max-gap clusters. We first present an exact expression for the probability of observing a complete max-gap cluster containing all m genes of interest within a randomly ordered genome of size n. We also provide an approximation for faster analysis. Next we extend this analysis to evaluate the probability of observing a cluster containing only a subset of the black genes. We present a simple dynamic programming algorithm that exactly calculates the probability of observing an incomplete cluster of size $h < m$, as well as an analytic solution for the case where $h > \frac{m}{2}$. We then use these equations to calculate the probability of clusters for a range of different genome sizes and parameter values. We discuss the influence of the parameters n, m, g and h on cluster significance and determine which regions of the parameter space yield clusters that are statistically significant. Finally we discuss some of the challenges that arise in extending this statistical model to whole-genome comparison.

2   Probabilities of Max-Gap Clusters

Our analytical tests of max-gap cluster significance are based on the probability of observing a cluster by chance in a genome with random gene order, the most basic null hypothesis we can consider. If we cannot reject that null hypothesis, no more complex, biologically motivated null hypothesis need be considered.

When calculating the probability of max-gap clusters it will be useful to know the number of ways of arranging m black genes to form a max-gap cluster within a window of length l. When both endpoints of the window contain a black gene the cluster will be of length *exactly* l and the problem is equivalent to a well-known sum-of-dice combinatorics problem [64]. Let

$$d_c(m, g, l) = \sum_{i=0}^{\lfloor (l-m)/(g+1) \rfloor} (-1)^i \binom{m-1}{i} \binom{l - i(g+1) - c}{m - c}.$$

When $c = 2$, $d_c(m, g, l)$ corresponds to the the number of ways of rolling $m - 1$ dice, each with faces numbered 0 to g, such that the sum of their faces is equal to $l - m$ [1]. This is equivalent to the number of ways of creating a max-gap cluster

---

[1] This in turn is equivalent to the number of ways of rolling a set of $m - 1$ dice, each of which has faces numbered 1 to $g + 1$, so that their cumulative sum in equal to $l - 1$, due to Uspensky [64].

of size m and length l since such a cluster has $m-1$ gaps with a cumulative sum of $l - m$.

The number of ways of generating a cluster with length no greater than l is equivalent to requiring that only one endpoint in the window contain a black gene. This is simply: $\sum_{r=m}^{l} d_2(m,g,r)$, which can be shown to be equivalent to $d_1(m,g,l)$ (see the Appendix for the derivation). Similarly, the number of ways of arranging m genes so that they form a max-gap cluster *anywhere* within a window of size l is $\sum_{r=m}^{l} d_1(m,g,r) = d_0(m,g,l)$.

These expressions will be used in the subsequent sections in various situations in which the length of a cluster is constrained. Note that an efficient implementation of d_c can be obtained by pre-computing all necessary factorials, allowing the entire summation to be computed in $O(l)$ time.

2.1 Exact and Approximate Probabilities for Complete Max-Gap Clusters

We begin by calculating the probability of observing a *complete* max-gap cluster. More formally, given a random genome of size n, what is the probability of observing all m black genes (in any order), such that the gap between adjacent black genes does not exceed g. We determine the probability by counting the number of ways to place all m genes in a genome of size n so that they form a max-gap cluster. We enumerate the clusters by the position of the leftmost black gene in the cluster. Given the position of the first black gene, there are $(g+1)^{m-1}$ ways to place the remaining black genes so that they form a max-gap cluster, which is simply the number of ways of choosing $m-1$ gaps so that the length of each gap is between 0 and g. The maximum possible length of a max-gap cluster is $w = m + g(m-1)$, and thus there are $n - w + 1$ ways of placing the first black gene so that a cluster of maximal length can be accommodated. In addition, the leftmost black gene could also be positioned within the $w-1$ genes at the end of the genome. The number of ways of placing m black genes to form a max-gap cluster in the last $w-1$ slots is precisely the quantity we derived in the previous section. Combining these terms, the probability of observing a complete max-gap cluster of m genes in a genome of size n is

$$P_M(n,m,g) = \frac{\max(0, n-w+1) \cdot (g+1)^{m-1} + d_0(m,g,\min(n,w-1))}{\binom{n}{m}}. \quad (1)$$

When $m \ll n$, the total number of permutations can be approximated in constant time using Stirling's approximation, and then the complexity of computing P_M is simply $O(w) = O(mg)$. Except when $w \geq n$, the running time will be independent of the genome size since the only calculation that is not constant time is computing the number of ways of constructing a max-gap cluster within the last $w-1$ genes in the genome. When a more efficient running time is required, we can construct a lower bound on the probability of observing a cluster by simply eliminating the final term that takes edge effects into account. We can compute an upper bound by instead assuming that all but the last $m-1$ positions in the genome can accommodate a cluster of maximal length:

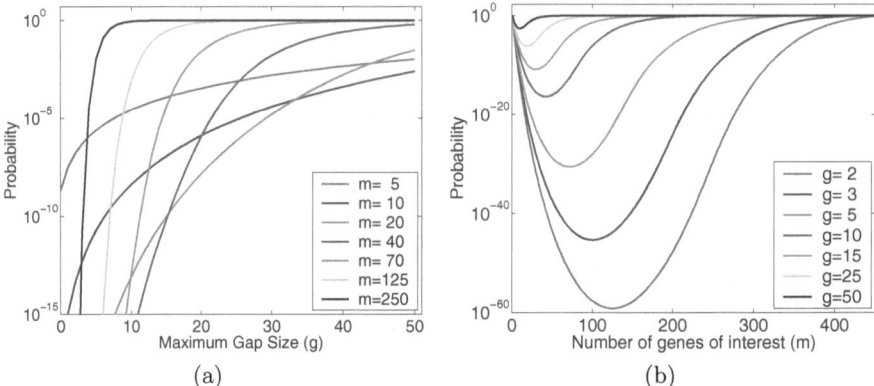

Fig. 2. Probability of a complete max-gap cluster of m black genes in a genome of size $n = 500$ as a function of g (a), and as a function of m (b).

$$\frac{\max(0, n - w + 1) \cdot (g+1)^{m-1}}{\binom{n}{m}} \leq P_M(n, m, g) \leq \frac{(n - m + 1) \cdot (g+1)^{m-1}}{\binom{n}{m}}.$$

Both bounds can be computed in constant time using Stirling's approximation to estimate the denominator. We have verified empirically that when n is large in relation to w, the upper bound is only a slight overestimate of P_M (data not shown).

In some cases we may wish to constrain the total length of the cluster, by adding the restriction that all m genes must appear in a window of size at most r. The limit on window size ensures a minimum cluster density, while the max-gap property prevents the gaps between black genes from becoming too large. More formally, given a genome of size n, the probability of finding all m black genes (in any order) in a window of size at most r such that the gap between adjacent black genes is never more than g, is simply

$$P_{MR}(n, m, g, r) = \frac{1}{\binom{n}{m}} \left[(n - r + 1) \cdot d_1(m, g, r) + d_0(m, g, r - 1) \right],$$

where we have replaced $(g+1)^{m-1}$ in Equation 1 with $d_1(m, g, r)$ in order to constrain the maximum length of the cluster.

The probability of finding a complete cluster for varying values of n, m, and g was calculated from Equation 1 using Mathematica. We selected parameter values corresponding to the range of values seen in real analyses. For example, we selected values of g ranging from 0 to 50, since typical values of this parameter used in genomic analyses range from three in bacteria [60] to about thirty in human [40]. We calculated probabilities for genomes sizes of 500, 1000, 5000, 20,000, and 25,000, corresponding to typical gene sets for bacteria, yeast, worm, and higher eukaryotes like human and *Arabidopsis*. For complete clusters we tested all values of m ranging from 2 to n.

Figure 2(a) shows the probability of observing a complete cluster containing all m black genes in a genome of size $n = 1000$, as m ranges from 1 to 250 and

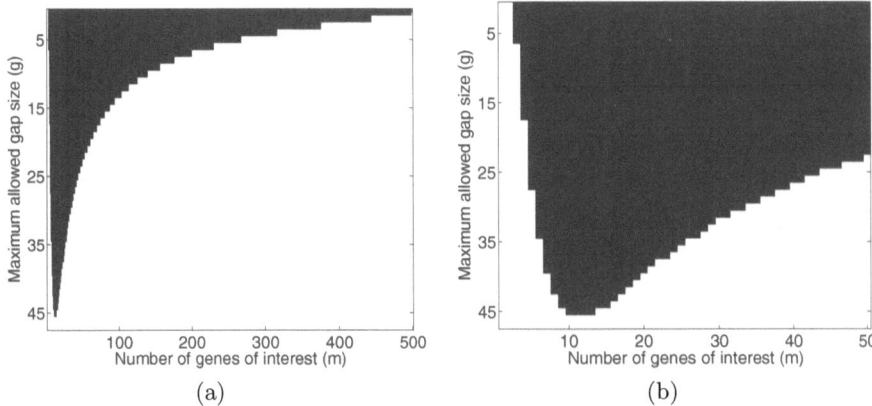

Fig. 3. Region of the parameter space that is statistically significant (shown in black) at the $\alpha = 0.0001$ level for a genome of size 500. (a) Complete parameter space where m ranges from 1 to 500. (b) Detail for $m \leq 50$.

g increases from 2 to 50. The probability of finding a complete cluster increases monotonically with g. We might also expect that this probability will increase monotonically with m, but this is not the case. As Figure 2(b) shows, as m increases, the probabilities first decrease and then increase. When m is small, a small increase in the number of black genes will actually decrease the probability of finding a cluster. This makes sense intuitively if ones considers the extreme cases: when $m = 1$ or $m = n$ the probability of finding a complete cluster will clearly be 1, and the values of m in between these two extremes will have probabilities of less than one.

One question of interest is the range of values of m and g for which is it possible to obtain a significant cluster. Figure 3 shows the parameter values for which the probability of observing a cluster in a genome of size 500 is no more than 0.0001. The significant region of the parameter space is shown in black, indicating that as gap size increases, the range of values of m for which it is possible to obtain a significant cluster becomes more and more restricted.

As the genome size n increases the probabilities decrease but the general trends seen in Figure 2 remain the same (data not shown).

2.2   Exact Probabilities for Incomplete Max-Gap Clusters

Requiring all m genes of interest to appear in a single cluster is often too strict a requirement. In practice, researchers often look for clusters that contain a subset of the genes of interest [1, 13, 19, 21, 30, 33, 34, 37, 39, 46, 48, 58, 59, 63]. Thus, we relax the cluster definition to allow incomplete clusters of size at least h, for $h < m$ (maintaining the requirement that there is no gap greater than g between adjacent black genes). Unlike complete clusters, there can be more than one incomplete cluster in the same genome. A simple extension of Equation 1 to incomplete clusters would therefore lead to overcounting permutations containing more than one cluster. Instead, we present a simple dynamic programming algorithm to count those permutations which *do not* contain a cluster of size h or larger, and subtract to obtain the probability of observing at least one incomplete cluster. The algorithm moves along the genome, adding a black or white gene at each step. It keeps track of runs of black genes that satisfy the max-gap cluster criterion and avoids creating a cluster of size h or larger by judicious placement of white genes.

The quantity $n_{\bar{H}}[n, m, j, c]$ represents the number of ways to place m black genes in n slots without creating a max-gap cluster of size greater than or equal to h, where j is the distance to the previous black gene and c is the size of any cluster created so far. It is defined recursively as follows:

$$
n_{\bar{H}}[n, m, j, c] = \begin{cases} 0, & \text{if } c = h \\ 0, & \text{else if } n < m \\ 1, & \text{else if } m = 0 \\ n_{\bar{H}}[n-1, m, j+1, c] + n_{\bar{H}}[n-1, m-1, 0, c+1], & \text{else if } j \leq g \\ n_{\bar{H}}[n-1, m, j+1, c] + n_{\bar{H}}[n-1, m-1, 0, 1], & \text{otherwise.} \end{cases}
$$

The probability of observing at least one incomplete cluster of size at least h is then just one minus the probability of containing no incomplete clusters

$$
P_H(n, m, h, g) = 1 - \frac{n_{\bar{H}}[n, m, g+1, 0]}{\binom{n}{m}}. \tag{2}
$$

The complexity of computing P_H is $O(nmgh)$. Since $h < m$, this is bounded above by $O(nm^2g)$. However, in practice m will be significantly smaller than n. For example, the size of typical bacterial genomes ranges from 500 to 5000 [57], whereas the average number of genes in an operon is predicted to be between two and four, and the large majority of operons contain fewer than fifteen genes [69]. Vertebrate genomes can be much larger. For example, the estimated size of the human genome is around 25,000 genes [32], but duplicated or conserved regions reported in the literature tend to include only five to thirty genes in a window containing a hundred genes at most [1, 13, 19, 21, 30, 33, 34, 37, 39, 46, 48, 58, 59, 63]. If we make the conservative assumption that $m \leq \sqrt{n}$ and that g is a small constant, then the running time will be bounded above by $O(n^2)$.

When $h > \frac{m}{2}$, the probability can be computed directly because we do not have to worry about overcounting genomes containing more than one cluster. We count the number of permutations containing a cluster, enumerating them by the position of the leftmost black gene in the leftmost cluster, just as we did for complete clusters. Unlike the complete case, however, we have to be careful not to overcount clusters of size greater than h. We accomplish this by considering each possible cluster length (for the first h black genes in the cluster) individually and placing $g + 1$ white genes before the start of the cluster to ensure that it cannot be extended to the left. This yields a probability of finding an incomplete cluster of size at least h of

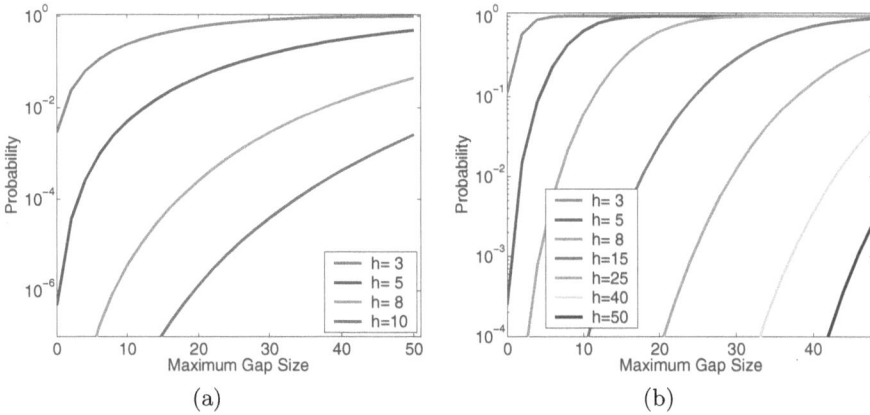

Fig. 4. Probability of an incomplete cluster of size at least h as a function of gap size in (a) a genome of 500 genes with $m = 10$ black genes, (b) a genome of 1000 genes with $m = 50$ black genes.

$$\frac{1}{\binom{n}{m}} \sum_{l=h}^{h+g(h-1)} \left[(n-l-g) \cdot d_0(h,g,l) \cdot \binom{n-l-g-1}{m-h} + E \right], \qquad (3)$$

where l ranges over all possible lengths of a cluster of size h and E is a term to address edge effects. The first term is the number of positions in which to start the cluster. The second term is the number of ways to choose the gaps to obtain a cluster length of exactly l. The third term is the number of ways to place the remaining $m - h$ genes outside the cluster. The final term counts clusters close to the beginning of the genome before which it is only possible to place $i < g+1$ white genes. It is calculated as

$$E = \sum_{i=0}^{g} d_0(h,g,l) \cdot \binom{n-l-i}{m-h} = d_0(h,g,l) \left[\binom{n-l+1}{m-h+1} - \binom{n-l-g}{m-h+1} \right],$$

where the binomials are defined to be zero when the upper value is smaller than the lower value and the simplification is by application of the upper summation identity [24]. The complexity of computing Equation 3 depends on the extent to which sub-computations are reused, but empirically we observe that even a naive implementation has a substantially faster running time than Equation 2 (data not shown).

We calculated the probability of finding an incomplete cluster from Equations 2 and 3 using Mathematica for the values of n and g given in Section 2.1. We chose to examine values of m ranging from 3 to 250, which covers the range of gene numbers found in typical reference regions of interest [1, 13, 19, 21, 30, 33, 34, 37, 39, 46, 48, 58, 59, 63], and values of h ranging from 3 to $m/2$. Figure 4 shows the probability of observing a cluster of a subset of 50 black genes in a genome of size 500 for varying values of g and h. As the maximum gap size

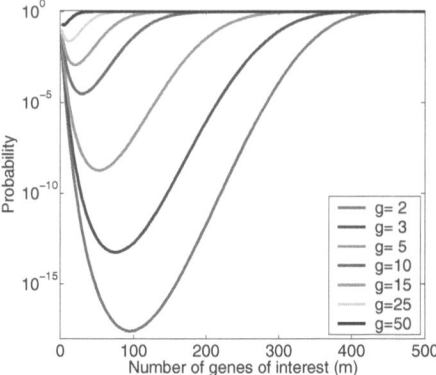

Fig. 5. Probability of observing a cluster that includes at least half of all m black genes in a genome of size 500.

allowed increases, so does the probability of finding an incomplete cluster. Increasing the required size (h) of the cluster, on the other hand, decreases its probability of occurring by chance. Figure 5 shows the probability of max-gap clusters for varying values of m, where $h = \frac{m}{2}$. As in the case of complete clusters, the probabilities first decrease then increase with m. Finally, Figure 6 shows the region of parameter space for which it is possible to find a significant cluster at a significance level of $\alpha = 0.0001$, when $m = 100$, for genomes of size $n = 500$ and $n = 1000$. Probabilities were also calculated for larger genome sizes as in Section 2.1. Again, as n increases the probabilities decrease but the general trends are similar (data not shown).

3   Discussion

The work presented here was motivated by the gap that currently exists between mathematical cluster models and models used in analysis of real genomes. We provide analytical statistical tests for max-gap clusters, a model widely used in practice [5, 10, 38, 40, 45, 60]. We determine the probability of observing a max-gap cluster containing a set of m pre-specified genes of interest, assuming a genome with random gene order. We also consider incomplete clusters, where a subset of the pre-specified genes satisfies the max-gap criterion. This scenario corresponds to a reference-region approach in which a particular chromosomal region in one genome is of interest, and another region containing a similar set of genes is sought. We have presented exact expressions for the probabilities of finding complete and incomplete max-gap clusters under this simple model. We have also provided an efficient approximation for the probability of finding a complete cluster, which is highly accurate when n is large in relation to mg.

Our calculations show that the probability of finding a cluster increases monotonically with g, and that as the gap size increases, the range of values of m for

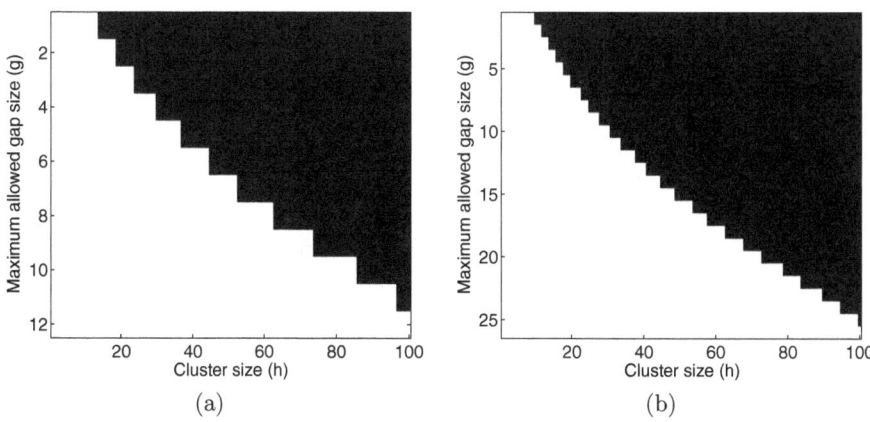

Fig. 6. Region of the parameter space that is statistically significant (shown in black) at the $\alpha = 0.0001$ level for $m = 100$ black genes in a genome of size $n = 500$ (a) and $n = 1000$ (b).

which it is possible to obtain a significant cluster becomes more and more restricted. For a fixed value of m, increasing the required size (h) of an incomplete cluster decreases its probability of occurring by chance. However, the behavior of cluster probabilities with respect to m is more complex. There is a high probability that all m black genes will form a cluster when m is small in relation to n, and this probability decreases as m grows larger. As m approaches n, however, the majority of genes in the genome will be black, and the probability that they cluster together begins to increase again. This behavior is also observed for incomplete clusters when h is chosen to be a fixed percentage of m.

The model considered here treats the genome as an ordered set of genes, disregarding actual distances between genes. This assumption can be advantageous because physical distances often differ substantially between organisms. Furthermore, it eliminates the need to model the variation in gene density that can lead to gene-rich and gene-poor regions of chromosomes. A distance-based model would have to take into account the fact that a cluster that is surprising in a gene-poor region might easily occur by chance in a gene rich region. However, since prokaryotic genomes tend to be gene dense, it would not be difficult to modify the model used here to a model that explicitly considers distance for bacteria. When analyzing clusters in bacterial genomes, statistical models that take into account the orientation of genes and the possibility of circular instead of linear chromosomes are also of interest. These extensions remain as future work.

The current model also disregards the presence of tandem duplications and gene families. Since tandem duplications can be detected easily in genomic data due to their regular spatial patterns, they can be taken into account by a preprocessing step in genomic analysis. Gene families are more problematic, however. Virtually all genomes contain gene families, sets of genes with similar sequence and function, that arose through duplication of genetic material. Large gene families will increase the likelihood of finding a conserved cluster by chance and, hence, can have a large impact on the statistical significance of a particular cluster. However, factoring gene families into an analytical statistical model is difficult because the exact size of each gene family in a genome cannot be easily determined.

An important open problem is the development of statistical tests for max-gap clusters in whole genome comparisons. More formally, given two genomes $G = (g_1,\ldots,g_n)$ and $H = (h_1,\ldots,h_n)$, and a mapping between homologs in G and H, we wish to find all maximal max-gap clusters containing at least k homologs.

It is not obvious how to calculate max-gap cluster probabilities in the case of whole-genome comparison because, unlike the abstraction of white and black genes presented here, in whole-genome comparison there is no specific set of genes that is of interest. Consider the simple model of whole genome comparison in which the genomes are assumed to have identical gene complements, and can therefore be treated as two permutations of the numbers $1,\ldots,n$. Although this model appears quite natural, max-gap clusters found under this approach to genome comparison have some surprising properties[2]:

1. Under this simple model of genome comparison with identical gene content, there will always be a cluster of size n and hence, the probability of finding a max-gap cluster of at least size k when comparing two genomes is always one. For example, consider these two genomes:

$$G = 1\ 2\ 3\ 4\ 5\ 6$$
$$H = 3\ 4\ 6\ 5\ 1\ 2$$

Suppose we wish to find the largest max-gap cluster that can be formed around gene 3, when $g = 0$. If we attempt to construct a cluster in a greedy fashion, the cluster will only include genes 3 and 4. However, if we look ahead a bit, it is possible to find the cluster [3 4 5 6]. In both genomes there are zero gaps between these four genes. Extending this look-ahead idea, we can see that under this model, regardless of the value of g, a pair of genomes always contain a max-gap cluster of size n. Since $n \geq k$, the probability of finding a cluster of size at least k is one.

2. In the reference region model discussed in this paper, as well as the gene cluster models of Durand and Sankoff [15] and Calabrese et al. [9], a cluster that contains k genes will always contain at least one valid cluster of size each from 1 to $k - 1$. However, this property does not hold when applying the max-gap cluster model to whole genome comparison. For example, consider the following two genomes:

$$G = \ldots\ 0\ \ 1\ \ 2\ \ 3\ \ 4\ \ 5\ \ 6\ \ 7\ \ 8\ \ 9\ 10\ 11\ 12\ \ldots$$
$$H = \ldots\ 2\ \ 4\ 27\ 30\ \ 9\ 12\ 53\ 81\ \ 0\ \ 8\ 99\ 72\ \ 7\ \ldots$$

---

[2] Bergeron and colleagues [4] have made similar observations in the context of the development of efficient algorithms for finding max-gap clusters, as opposed to the statistical questions considered here.

With a maximum allowed gap of $g = 2$, the size of the largest max-gap cluster is seven: [0 2 4 7 8 9 12]. However, this cluster does not contain any valid max-gap clusters of size three to six. Indeed, it contains only sub-clusters of size two ([2 4], [9 12], and [7 8]).

This issue is related to point (1). There may be a higher probability of finding a larger cluster than a smaller cluster. To see why this is the case, note that increasing the size of the cluster essentially increases the maximum allowed window size. As a result, as the size of the cluster sought increases, the number of clusters found may grow substantially.

When looking for evidence of whole-genome duplication, a genome is compared with itself, and the gene sets will indeed be identical. In the comparison of two different genomes, however, point (1) will not be an issue, because gene sets are never identical in practice. This problem can be partially addressed by a more realistic model, where only a subset of the gene sets of the two genomes are shared. We assume that only m genes in each genome have homologs in the other genome, and the non-homologous genes are randomly distributed throughout the genome. When $g = 0$, the non-matching genes will create a natural barrier to unlimited extension of a cluster, preventing the formation of a max-gap cluster of size m. However, if g is greater than the longest contiguous run of non-matching genes then it will still be possible to form a cluster of size m.

Furthermore, this more realistic model does not circumvent the second issue of non-monotonic cluster sizes. These two issues have implications for the development of analytical statistical models of max-gap clusters found through whole-genome comparison, and remain exciting problems for the future.

Acknowledgments

D.D. was supported by NIH grant 1 K22 HG 02451-01 and a David and Lucille Packard Foundation fellowship. D.S. was supported in part by grants from the Natural Sciences and Engineering Research Council of Canada. He holds the Canada Research Chair in Mathematical Genomics and is a Fellow of the Evolutionary Biology Program of the Canadian Institute for Advanced Research. R.H. was supported in part by a Barbara Lazarus Women@IT Fellowship, funded in part by the Alfred P. Sloan Foundation.

References

1. A. Amores, A. Force, Y. l. Yan, L. Joly, C. Amemiya, A. Fritz, R.K. Ho, J. Langeland, V. Prince, Y. L. Wang, M. Westerfield, M. Ekker, and J. H. Postlethwait. Zebrafish hox clusters and vertebrate genome evolution. *Science*, 282:1711–1714, 1998.
2. Arabidopsis Genome Initiative. Analysis of the genome sequence of the flowering plant *Arabidopsis thaliana*. *Nature*, 408:796–815, 2000.
3. A. K. Bansal. An automated comparative analysis of 17 complete microbial genomes. *Bioinformatics*, 15:900–908, 1999.

4. A. Bergeron, S. Corteel, and M. Raffinot. The algorithmic of gene teams. In D. Gusfield and R. Guigo, editors, *Algorithms in Bioinformatics, Second International Workshop WABI2002*, Lecture Notes in Computer Science 2452, pages 464–476, 2002.

5. G. Blanc, K. Hokamp, and K.H. Wolfe. A recent polyploidy superimposed on older large-scale duplications in the arabidopsis genome. *Genome Res*, 13(2):137–44, 2003.

6. M. Blanchette, T. Kunisawa, and D. Sankoff. Gene order breakpoint evidence in animal mitochondrial phylogeny. *Journal of Molecular Evolution*, 49:193–203, 1999.

7. P Bork, B. Snel, G. Lehmann, M. Suyama, T. Dandekar, W. Lathe III, and M. Huynen. Comparative genome analysis: exploiting the context of genes to infer evolution and predict function. In D. Sankoff and J. H. Nadeau, editors, *Comparative Genomics*, pages 281–294. Kluwer Academic Press, Dordrecht, NL, 2000.

8. G. Bourque and P.A. Pevzner. Genome-scale evolution: Reconstructing gene orders in the ancestral species. *Genome Res*, 12(1):26–36, 2002.

9. P. P. Calabrese, S. Chakravarty, and T. J. Vision. Fast identification and statistical evaluation of segmental homologies in comparative maps. *ISMB (Supplement of Bioinformatics)*, pages 74–80, 2003.

10. X Chen, Z Su, P Dam, B Palenik, Y Xu, and T Jiang. Operon prediction by comparative genomics: an application to the Synechococcus sp. WH8102 genome. *Nucleic Acids Res*, 32(7):2147–2157, 2004.

11. A. Coghlan and K. H. Wolfe. Fourfold faster rate of genome rearrangement in nematodes than in *Drosophila*. *Genome Research*, 12(6):857–867, 2002.

12. M. E. Cosner, R. K. Jansen, B. M. E. Moret, L. A. Raubeson, L.-S. Wang, T. Warnow, and S. Wyman. An empirical comparison of phylogenetic methods on chloroplast gene order data in *Campanulaceae*. In D. Sankoff and J. H. Nadeau, editors, *Comparative Genomics*, pages 99–121. Kluwer Academic Press, Dordrecht, NL, 2000.

13. F. Coulier, P. Pontarotti, R. Roubin, H. Hartung, M. Goldfarb, and D. Birnbaum. Of worms and men: An evolutionary perspective on the fibroblast growth factor (FGF) and FGF receptor families. *J. Mol Evol*, 44:43–56, 1997.

14. E.G. Danchin, L.Abi-Rached, A. Gilles, and P. Pontarotti. Abstract conservation of the mhc-like region throughout evolution. *Immunogenetics*, 5(3):141–8, 2003.

15. D. Durand and D. Sankoff. Tests for gene clustering. *Journal of Computational Biology*, 10(3/4):453–482, 2003.

16. J. Ehrlich, D. Sankoff, and J.H. Nadeau. Synteny conservation and chromosome rearrangements during mammalian evolution. *Genetics*, 147(1):289–96, 1997.

17. N. El-Mabrouk, J. H. Nadeau, and D. Sankoff. Genome halving. In Springer-Verlag, editor, *Combinatorial Pattern Matching*, pages 235–250, 1998.

18. N. El-Mabrouk and D. Sankoff. The reconstruction of doubled genomes. *SIAM Journal of Computing*, 32:754–792, 2003.

19. T. Endo, T. Imanishi, T. Gojobori, and H. Inoko. Evolutionary significance of intra-genome duplications on human chromosomes. *Gene*, 205(1–2):19–27, 1997.

20. M. D. Ermolaeva, O. White, and S. Salzberg. Prediction of operons in microbial genomes. *Nucleic Acids Res*, 5(29):1216–1221, Mar 2001.

21. T.J. Gibson and J. Spring. Evidence in favour of ancient octaploidy in the vertebrate genome. *Biochem Soc Trans*, 2:259–264, Feb 2000.

22. D. Goldberg, S. McCouch, and J. Kleinberg. Algorithms for constructing comparative maps. In D. Sankoff and J. H. Nadeau, editors, *Comparative Genomics*, pages 281–294. Kluwer Academic Press, Dordrecht, NL, 2000.

23. L. A. Goldberg, P. W. Goldberg, M. S. Paterson, P. Pevzner, S. C. Sahinalp, and E. Sweedyk. The complexity of gene placement. *Journal of Algorithms*, 41(2):225–2435, 2001.
24. Graham, Knuth, and Patashnik. *Concrete Mathematics*. Addison-Wesley, 1989.
25. S. Hampson, A. McLysaght, B. Gaut, and P. Baldi. LineUp: statistical detection of chromosomal homology with application to plant comparative genomics. *Genome Res*, 13(5):999–1010, 2003.
26. S. Hannenhalli, C. Chappey, E. V. Koonin, and P. A. Pevzner. Genome sequence comparison and scenarios for gene rearrangements: A test case. *Genomics*, 30:299 – 311, 1995.
27. S. Heber and J. Stoye. Algorithms for finding gene clusters. In *Proceedings of WABI01*, Lecture Notes in Computer Science 2149, pages 254–265, 2001.
28. S. Heber and J. Stoye. Finding all common intervals of k permutations. In *Proceedings of CPM01*, Lecture Notes in Computer Science 2089, pages 207–218, 2001.
29. E. A. Housworth and J. Postlethwait. Measures of synteny conservation between species pairs. *Genetics*, 162(1):441–8, 2002.
30. A. L. Hughes. Phylogenetic tests of the hypothesis of block duplication of homologous genes on human chromosomes 6, 9, and 1. *MBE*, 15(7):854–70, 1998.
31. M. Huynen and P. Bork. Measuring genome evolution. *Proc Natl Acad Sci U S A*, 95:5849–56, 1998.
32. International Human Genome Sequencing Consortium. Initial sequencing and analysis of the human genome. *Nature*, 409(682):860–921, 2001.
33. M. Kasahara. New insights into the genomic organization and origin of the major histocompatibility complex: role of chromosomal (genome) duplication in the emergence of the adaptive immune system. *Hereditas*, 127(1–2):59–65, 1997.
34. N. Katsanis, J. Fitzgibbon, and E.M. Fisher. Paralogy mapping: identification of a region in the human MHC triplicated onto human chromosomes 1 and 9 allows the prediction and isolation of novel PBX and NOTCH loci. *Genomics*, 35(1):101–8, 1996.
35. A. B. Kolsto. Dynamic bacterial genome organization. *Molecular Microbiology*, 24:241–8, 1997.
36. J.G. Lawrence and J. R. Roth. Selfish operons: horizontal transfer may drive the evolution of gene clusters. *Genetics*, 143:1843–60, 1996.
37. L. Lipovich, E. D. Lynch, M. K. Lee, and M-C. King. A novel sodium bicarbonate cotransporter-like gene in an ancient duplicated region: *SLC4A9* at 5q31. *Genome Biology*, 2(4):0011.1–0011.13, 2001.
38. N. Luc, J.L. Risler, A. Bergeron, and M. Raffinot. Gene teams: a new formalization of gene clusters for comparative genomics. *Comput Biol Chem.*, 27(1):59–67, 2003.
39. L. G. Lundin. Evolution of the vertebrate genome as reflected in paralogous chromosomal regions in man and the house mouse. *Genomics*, 16(1):1–19, 1993.
40. A. McLysaght, K. Hokamp, and K. H. Wolfe. Extensive genomic duplication during early chordate evolution. *Nat Genet.*, 31(2):200–204, 2002.
41. J. H. Nadeau and B. A. Taylor. Lengths of chromosomal segments conserved since the divergence of man and mouse. *Proc.Natl.Acad.Sci. USA*, 81:814–818, 1984.
42. J.H. Nadeau and D. Sankoff. Counting on comparative maps. *Trends Genet*, 14(12):495–501, 1998.
43. J.H. Nadeau and D. Sankoff. The lengths of undiscovered conserved segments in comparative maps. *Mamm Genome*, 9(6):491–5, 1998.
44. S. J. O'Brien, J. Wienberg, and L. A. Lyons. Comparative genomics: lessons from cats. *Trends Genet*, 10(13):393–399, Oct 1997.

45. R. Overbeek, M. Fonstein, M. D'Souza, G. D. Pusch, and N. Maltsev. The use of gene clusters to infer functional coupling. *PNAS*, 96:2896–2901, 1999.

46. M.-J. Pebusque, F. Coulier, D. Birnbaum, and P. Pontarotti. Ancient large-scale genome duplications: phylogenetic and linkage analyses shed light on chordate genome evolution. *MBE*, 15(9):1145–59, 1998.

47. Pavel A. Pevzner. *Computational Molecular Biology: An Algorithmic Approach.* MIT Press, Cambridge, MA, 2000.

48. I. Ruvinsky and L. M. Silver. Newly indentified paralogous groups on mouse chromosomes 5 and 11 reveal the age of a t-box cluster duplication. *Genomics*, 40:262–266, 1997.

49. D. Sankoff, D. Bryant, M. Deneault, B. F. Lang, and G. Burger. Early eukaryote evolution based on mitochondrial gene order breakpoints. *J Comput Biol*, 3–4:521–535, 2000.

50. D. Sankoff, M. Deneault, D. Bryant, C. Lemieux, and M. Turmel. Chloroplast gene order and the divergence of plants and algae from the normalized number of induced breakpoints. In D. Sankoff and J. H. Nadeau, editors, *Comparative Genomics*, pages 89–98. Kluwer Academic Press, Dordrecht, NL, 2000.

51. D Sankoff and N. El-Mabrouk. Genome rearrangement. In T. Jiang, T. Smith, Y. Xu, and M. Zhang, editors, *Current Topics in Computational Biology*, pages 135–155. MIT Press, 2002.

52. D. Sankoff, V. Ferretti, and J. H. Nadeau. Conserved segment identification. *Journal of Computational Biology*, 4:559–565, 1997.

53. C. Semple and K. H. Wolfe. Gene duplication and gene conversion in the *Caenorhabditis elegans* genome. *JME*, 48(5):555–64, 1999.

54. C. Seoighe and K. H. Wolfe. Updated map of duplicated regions in the yeast genome. *Gene*, 238:253–261, 1999.

55. C. Seoighe and K.H. Wolfe. Extent of genomic rearrangement after genome duplication in yeast. *Proc Natl Acad Sci U S A*, 95(8):4447–52, 1998.

56. C. Simillion, K. Vandepoele, M.C. Van Montagu, M. Zabeau, and Y. Van de Peer. The hidden duplication past of arabidopsis thaliana. *Proc Natl Acad Sci U S A*, 99(21), 2002.

57. M Skovgaard, L J Jensen, S Brunak, D Ussery, and A Krogh. On the total number of genes and their length distribution in complete microbial genomes. *Trends Genet*, 17(8):425–428, Aug 2001.

58. N. G. C. Smith, R. Knight, and L. D. Hurst. Vertebrate genome evolution: a slow shuffle or a big bang. *BioEssays*, 21:697–703, 1999.

59. J. Spring. Genome duplication strikes back. *Nature Genetics*, 31:128–129, 2002.

60. J. Tamames. Evolution of gene order conservation in prokaryotes. *Genome Biol*, 6(2):0020.1–11, 2001.

61. J. Tamames, G. Casari, C. Ouzounis, and A. Valencia. Conserved clusters of functionally related genes in two bacterial genomes. *JME*, 44::66–73, 1997.

62. J. Tamames, M. Gonzalez-Moreno, A. Valencia, and M. Vicente. Bringing gene order into bacterial shape. *Trends Genet*, 3(17):124–126, Mar 2001.

63. Z. Trachtulec and J. Forejt. Synteny of orthologous genes conserved in mammals, snake, fly, nematode, and fission yeast. *Mamm Genome*, 3(12):227–231, Mar 2001.

64. J. V. Uspensky. *Introduction to Mathematical Probability*, pages 23–24. McGraw-Hill, New York, 1937.

65. K. Vandepoele, Y. Saeys, C. Simillion, J. Raes, and Y. Van De Peer. The automatic detection of homologous regions (ADHoRe) and its application to microcolinearity between arabidopsis and rice. *Genome Res*, 12(11):1792–801, 2002.

66. J. C. Venter et al. The sequence of the human genome. *Science*, 291(5507):1304–51, 2001.
67. T. J. Vision, D. G. Brown, and S. D. Tanksley. The origins of genomic duplications in Arabidopsis. *Science*, 290:2114–2117, 2000.
68. K. H. Wolfe and D. C. Shields. Molecular evidence for an ancient duplication of the entire yeast genome. *Nature*, 387:708–713, 1997.
69. Yu Zheng, Joseph D Szustakowski, Lance Fortnow, Richard J Roberts, and Simon Kasif. Computational identification of operons in microbial genomes. *Genome Res*, 12(8):1221–1230, Aug 2002.

A   Derivation of $d_1(m, g, r)$ from $d_2(m, g, r)$

In Section 2 we gave an expression $d_2(m, g, l)$ for the number of ways of arranging m black genes into a max-gap cluster of length *exactly* l.

The number of ways $d_1(m, g, l)$ of arranging m black genes in a max-gap cluster of length *no greater* than l is as follows:

$$\sum_{r=m}^{l} d_2(m, g, r) = \sum_{r=m}^{l} \sum_{i=0}^{\lfloor (r-m)/(g+1) \rfloor} (-1)^i \binom{m-1}{i} \binom{r - i(g+1) - 2}{m - 2},$$

The r in the upper bound of the second summation can be replaced by l because when $i > \lfloor (r-m)/(g+1) \rfloor$ the final binomial will be zero, which gives

$$\sum_{r=m}^{l} \sum_{i=0}^{\lfloor (l-m)/(g+1) \rfloor} (-1)^i \binom{m-1}{i} \binom{r - i(g+1) - 2}{m - 2}.$$

Now the upper bound of the second summation is no longer dependent on r, and so the outer summation can be moved inward:

$$\sum_{i=0}^{\lfloor (l-m)/(g+1) \rfloor} (-1)^i \binom{m-1}{i} \sum_{r=m}^{l} \binom{r - i(g+1) - 2}{m - 2}.$$

Rewriting the bounds of the inner summation gives:

$$\sum_{i=0}^{\lfloor (l-m)/(g+1) \rfloor} (-1)^i \binom{m-1}{i} \sum_{r=m-i(g+1)-2}^{l-i(g+1)-2} \binom{r}{m - 2}.$$

Decreasing the lower bound to $r = 0$ does not affect the probability because when $0 \leq r < m - 2$ the binomial is zero. We apply the upper summation identity [24] to eliminate the inner summation, which yields

$$\sum_{i=0}^{\lfloor (l-m)/(g+1) \rfloor} (-1)^i \binom{m-1}{i} \binom{l - i(g+1) - 1}{m - 1},$$

which is exactly $d_1(m, g, r)$. The derivation of $d_0(m, g, r)$ from $d_1(m, g, r)$ is identical.

Identifying Evolutionarily Conserved Segments Among Multiple Divergent and Rearranged Genomes

Bob Mau[1,2,*], Aaron E. Darling[1,3], and Nicole T. Perna[1,4]

[1] Dept. of Animal Health and Biomedical Sciences,
University of Wisconsin - Madison, 1656 Linden Dr., Madison, WI 53706, USA
[2] Dept. of Oncology, University of Wisconsin - Madison
1656 Linden Dr., Madison, WI 53706, USA
[3] Dept. of Computer Science, University of Wisconsin - Madison
1656 Linden Dr., Madison, WI 53706, USA
[4] The Genome Center of Wisconsin
University of Wisconsin - Madison
1656 Linden Dr., Madison, WI 53706, USA

Abstract. We describe a new method for reliably identifying conserved segments among genome sequences that have undergone rearrangement, horizontal transfer, and substantial nucleotide-level divergence. A Gibbs-like sampler explores different combinations of sequence-based markers shared by the genomes under study. The sampler assigns each marker a posterior probability based on how frequently it participates in some collinear group of markers. Markers with high p.p. values are likely members of conserved segments. The method identifies both large-scale and local trends in segmental collinearity, providing suitable input for genome alignment and rearrangement history inference tools. Applying our method to genomes of four *Streptococci* reveals that rearranged segments in these organisms belong in two size categories: large conserved segments that are interrupted by a staccato of single gene or operon-size small segments. The rearrangement pattern of large segments is best explained by symmetric inversions about the origin of replication while the pattern of small segments is not.

1 Introduction

Nadeau and Taylor [1] introduced the concept of 'conserved segments' when comparing the genetic linkage maps of human and mouse. Conserved segments are homologous regions between genomes in which common genetic markers occur in the same order. Twenty years later, comparison of the completed human and mouse genomes found one third more large-scale rearrangements than predicted [2], as well as thousands of micro-rearrangements [3]. Similarly, the discovery of a large inversion between *E. coli* and *Salmonella typhimurium* by genetic analysis [4] spurred studies of genomic rearrangements in microbes. Pairwise comparisons of sequenced eubacterial genomes later confirmed that symmetric

---

* To whom correspondence should be addressed: `robertm@genome.wisc.edu`

J. Lagergren (Ed.): RECOMB 2004 Ws on Comparative Genomics, LNBI 3388, pp. 72–84, 2005.
© Springer-Verlag Berlin Heidelberg 2005

inversions about the origin and terminus of replication are common in this domain [5, 6]. Such pairwise comparisons are easy to implement but provide only limited analytical power, whereas multiple genome comparison enables the application of more powerful phylogenetic methods.

We present a method to reliably identify conserved blocks of sequence among several genomes that have undergone rearrangement. Our approach to rearrangement identification relies on monotypic markers to suggest potential homology. Monotypic markers are genomic features that occur exactly once in each genome being compared. The order and orientation in which these markers appear can be written as signed permutations of integers. Applying breakpoint analysis [7] to these permutations separates the markers into disjoint subsets of collinear markers. The regions of DNA spanned by the markers inside a given subset form a locally collinear block, or LCB. Unlike the conserved segments originally described by Nadeau and Taylor, LCBs are based solely on sequence similarity and do not imply any type of common evolutionary history or biological significance. In particular, LCBs make no distinction between segments that are similar merely by chance and truly orthologous segments - segments whose similarity derives from a single locus in the most recent common ancestor (MRCA). Conserved segments are regions of strictly orthologous sequences [8] that may contain lineage specific sequence, but do not contain rearrangements of orthologous sequence. By using marker order rather than the chromosomal proximity of markers to assess segmental conservation, our method accommodates lineage specific lateral gene transfer. Previous analytical tools were either limited to closely-related taxa [9] or did not account for horizontal transfer events common in bacterial genomes [3, 10].

Our target data set for this work consists of a group of four *Streptococcus* species that have sufficiently diverged so that comparisons at the nucleotide level are not practicable. We have shown in earlier studies [9, 11] that multi-MUMs (multiple maximum unique matches) are simple and effective monotypic markers for finding genomic rearrangements, but their applicability is limited to closely related organisms. When comparing more distantly related genomes, exact nucleotide matches such as multi-MUMs fail to generate a sufficiently comprehensive set of monotypic markers. Various types of inexact matching algorithms have been designed with DNA sequence in mind [12–15], but BLAST [16] hits at the protein level remain one of the most sensitive and widely used pairwise alternatives. In order to compare three or more annotated genomes, we define a gene-based monotypic marker to consist of a single gene from each genome, where each gene is the reciprocal best BLAST hit of the other genes comprising the marker. Although our focus here is on gene-based marker sets, we present the algorithm in full generality.

2   Notation

We start with a collection of G genomes $\mathcal{G}$ and M sequence-based markers $\mathcal{M}$ such that each marker occurs once and only once in each $g \in \mathcal{G}$. One genome, denoted g_1, is designated as a reference genome. In order to facilitate breakpoint determination, we assign an integer label to each marker based on the marker's order in g_1 coordinates, *i.e.* the jth marker in g_1 is labelled j. In other genomes, m_j may reside on the opposite strand, in which case that instance is labelled $-j$. Hence, $m_j = \pm j$, depending on its orientation in g_k relative to g_1. Denote the G labels of m_j by $\phi_k(m_j) = \pm j$. A signed permutation ζ_k is constructed for each genome by sorting the M integers of $\phi_k(\mathcal{M})$ by their location in g_k. ζ_k thereby encodes the order and relative orientation of monotypic markers in g_k. Throughout, markers m are indexed by j or v, genomes g by k, and integer elements z in ζ_k will be indexed by i. When it is clear from context, the marker m_j is denoted by its label j for ease of exposition.

An adjacency in $\zeta = (z_1, ..., z_i, ..., z_M)$ exists whenever $z_{i+1} - z_i = 1$. Conversely, $z_{i+1} - z_i \neq 1$ indicates a breakpoint between markers i and $i+1$. Breakpoints in ζ partition $\mathcal{M}$ into locally collinear blocks, groups of consecutively increasing integers. For example, $\zeta = (1, 2, 3, 4, 5, -8, -7, -6, 10, 9, -13, -12, -11, 14)$ consists of six collinear blocks: $\{1, 2, 3, 4, 5\}, \{6, 7, 8\}, \{11, 12, 13\}$, and three singletons: $\{9\}, \{10\}$, and $\{14\}$. Extension to three or more genomes is carried out by stacking permutations on top of one another to create a G x M matrix of integer-valued permutation elements: $\mathbf{Z}_G(\mathcal{M}) = (\zeta_1, ..., \zeta_G)$. The breakpoints of $\mathbf{Z}_G(\mathcal{M})$ are the union of breakpoints from each row ζ_k. In this framework, locally collinear blocks can be viewed as groups of consecutive integers that are present in either orientation in every row.

Our MCMC algorithm utilizes this matrix representation to determine LCBs "on the fly". After locating the G instances of m in $\mathbf{Z}_G(\mathcal{M})$, the algorithm quickly identifies all surrounding collinear markers by scanning to the left and right of m in all rows (genomes) simultaneously until the first breakpoint is encountered in each direction.

The procedure can be formalized as follows. Let $z_k(m_j)$ be the unique occurrence of m_j in ζ_k. If $i(m_j, k)$ denotes the relative position of marker m_j in g_k (equivalently, in ζ_k), then $z_k(m_j) = z(k, i(m_j, k))$ in $\mathbf{Z}_G(\mathcal{M})$. In particular, $z_1(m_j) = z(1, j) = j$. Define a shift operator θ on $z_k(m)$ by $\theta^h(z_k(m)) = z(k, i(m, k) + \text{sgn}(\phi_k(m)* h)$. Hence, $\theta^h(z_k(m))$ is the hth marker to the right or left of m in ζ_k. For h>0, shift h units to the right in ζ_k whenever m is located on the same strand (*i.e.*, $\phi_k(m) = 1$) and h units to the left when m occurs on the opposite strand. Reverse directions for h<0. Trivially, $\theta^0(z_k(m)) = z_k(m)$. The limits of collinearity about marker $m = m_j$ can be formulated as:

$$\text{Lend}(m : \mathbf{Z}(\mathcal{M})) = j + \min_{u \leq 0}\{j + h = \text{sgn}(\phi_k(m))\theta^h(z_k(m)) \forall k, h : 0 \geq h \geq u\}$$
$$\text{Rend}(m : \mathbf{Z}(\mathcal{M})) = j + \max_{u \geq 0}\{j + h = \text{sgn}(\phi_k(m))\theta^h(z_k(m)) \forall k, h : 0 \leq h \leq u\}$$

In a sense, this can be viewed as a seed and extend method to identify LCBs. The set of markers $\{m_v : \text{Lend}(m_j : \mathbf{Z}_G(\mathcal{M})) \leq v \leq \text{Rend}(m_j : \mathbf{Z}_G(\mathcal{M}))\}$ comprises the longest uninterrupted stretch of collinear, co-oriented markers containing m. The need for explicit recognition of the dependence on $\mathbf{Z}_G(\mathcal{M})$ will become clear shortly.

First, we give an example consisting of fourteen markers in four genomes and find the LCB containing m_7.

$$\mathbf{Z}_G(\mathcal{M}) = \begin{pmatrix} 1 & 2 & 3 & 4 & 5 & 6 & 7 & 8 & 9 & 10 & 11 & 12 & 13 & 14 \\ 1 & 2 & 3 & 4 & 9 & -8 & -7 & -6 & -5 & 10 & -13 & -12 & -11 & 14 \\ 1 & 2 & -5 & -4 & -3 & 14 & 6 & 7 & 8 & 9 & 11 & 12 & 13 & -10 \\ 6 & 7 & 8 & 9 & 1 & 2 & 3 & 4 & 5 & 10 & 11 & 12 & 13 & 14 \end{pmatrix}$$

For $m = m_7$, $z_1(m) = 7$, $z_2(m) = -7$, $i(7,2) = 7$, $z_3(m) = +7$, $i(7,3) = 8$, $z_4(4) = +7$ and $i(7,4) = 2$. The first mismatch to the right of marker 7 occurs two markers to the left of -7 in g_2, where $\mathrm{sgn}(\phi_2(m)) \times \theta^2(z_2(m) = -7) = -1 \times 9 \neq 9$. Hence, $\mathrm{Rend}(7) = 8$. Likewise, breakpoints in g_3 and g_4 occur two markers to the left of 7, so $\mathrm{Lend}(7) = 6$. Consequently, $\mathrm{LCB}(7) = \{6,7,8\}$.

3   A Pseudo-Gibbs Sampler

Not all markers help define evolutionarily conserved segments among genomes. On the contrary, some markers actually disrupt such segments. The problem is to identify those markers that optimally segregate into conserved segments. We attack the problem by surveying and assessing candidate subsets using Markov chain Monte Carlo technology. Each subset of $\mathcal{M}$ can be represented by a vector of M zeroes and ones, where a one in the jth position indicates that the jth marker is included in the subset, denoted by $\mathcal{M}^{in}$. We call this representation a configuration. In the previous example, the configuration $(0,0,0,0,0,1,1,1,0,1,1,1,0)$ corresponds to the subset $\mathcal{M}^{in} = \{6,7,8,11,12,13\}$. Denote the collection of locally collinear blocks of $\mathcal{M}^{in}$ by $\mathcal{L}(\mathcal{M}^{in})$ to emphasize the dependence on the set of included markers $\mathcal{M}^{in}$. Here, $\mathcal{L}(\mathcal{M}^{in})$ consists of two blocks: (6,7,8) and (11,12,13). Two different configurations on $\mathcal{M}$ define distinct inclusion subsets $\mathcal{M}_1^{in}$ and $\mathcal{M}_2^{in}$. $\mathcal{L}(\mathcal{M}_1^{in})$ and $\mathcal{L}(\mathcal{M}_2^{in})$ are considered equal if their LCBs span the same intervals in every genome (cf. $\mathcal{M}^{in} = \{6,8,11,13\}$ for configuration $(0,0,0,0,0,1,0,1,0,1,0,1,0)$).

We designed a pseudo-Gibbs sampler to explore the space of configurations in search of well-supported subsets as follows. Assign random variables X_j that map each m_j into a state of inclusion (1) or exclusion (0). Hence, the random vector $\mathbf{X}(\mathcal{M}) = (X_1(m_1),\ldots,X_M(m_M))$ maps $\mathcal{M}$ into $\mathbf{x} = (x_1,\ldots,x_M)$, a configuration of size M. Initialize $\mathbf{X}^0(\mathcal{M}) = \mathbf{x}^0$ with a draw of M independent Bernoulli($\frac{1}{2}$) random variates. A Markov chain $(\mathbf{X}^0, \mathbf{X}^1, \ldots, \mathbf{X}^n, \ldots, \mathbf{X}^N)$ is run over the space of configurations as follows. Pick a marker at random and compute a score conditioned on the current configuration $\mathbf{X}^n = \mathbf{x}$. Then convert the score to a "conditional probability" to stochastically update $X_j^{n+1}(m_j)$. The specific formulae used are:

$$\mathrm{Score}(m_j \mid \mathbf{X}^n = \mathbf{x}) = \sum_{v=L}^{j-1} w_v x_v + \max(\lambda(w_j - w_{min}), 0) + \sum_{v=j+1}^{R} w_v x_v, \quad (1)$$

where L $= \mathrm{Lend}(m : \mathbf{Z}_G(\mathcal{M}_n^{in}))$ and R $= \mathrm{Rend}(m : \mathbf{Z}_G(\mathcal{M}_n^{in})$ as defined above, w_m is a marker's weight, and λ and w_{min} down-weight the current marker.

$$\widehat{p}_j = \frac{e^{Score(m_j)/c} - 1}{e^{Score(m_j)/c} + 1}, \text{where } c > 0 \text{ is a scale parameter.} \quad (2)$$

$$\text{Sample } u \sim \text{Unif } [0,1]. \quad \text{If } u \geq \widehat{p}_j, \quad X_j^{n+1} = 0, \text{ else } X_j^{n+1} = 1. \qquad (3)$$

The score in (1) is the sum of the weights of all the collinear markers to the left and the right of m in the current configuration $\mathbf{x}$. When sets of markers consist of exact sequence matches, the weight w_m is simply the length of the match. For gene-based markers, calculation of w_m is complicated because every pair of reciprocal best BLAST hits generates a different BLAST bit score. We compute a gene-based marker's weight w_m as the square root of the average bit score over all possible genome pairs. The square root transformation reduces the distributional skew of large scores in long genes. The formula for the update probability in (2) is the right half of a sigmoidal function.

For the analysis described here, LCBs consisting of one or two genes are not particularly illuminating. In the case of phylogenetic reconstruction based on rearrangements, they can lead to false inferences. Rather than summarily exclude such blocks, their frequency can be minimized by down-weighting the current marker in the score function. Subtracting a minimum weight offset w_{min} suffices for nucleotide based markers, but with gene-based markers, an additional multiplicative reduction is required (*i.e.* $\lambda < 1$).

The pseudo-Gibbs sampler iterates through these three steps tens of millions of times. From a random start $\mathbf{x_0}$, the chain undergoes a burn-in period before entering well-supported configurations. These early realizations are discarded. As the Markov chain converges, some markers coalesce into collinear blocks because markers within a collinear block contribute to each other's scores. Counters record the number of times each marker is updated and the number of times the update is a one.

When the Markov chain has completed its pre-assigned number of iterations, the relative frequencies of inclusion in $\mathcal{M}^{in}$ are computed from the recorded counts for each m. Had we a bona fide Gibbs sampler, in which the conditional distributions were consistent with some joint probability distribution, the Hammersley-Clifford theorem [17] and standard Markov chain Monte Carlo theory [18] would guarantee that the relative frequency at each node converges to the appropriate marginal posterior probability.

$$\frac{\# \text{ of ones at } m_j}{\# \text{ of visits to } m_j} \rightarrow \pi_j(c) = \text{Pr}(m_j \text{ present in } \mathcal{M}^{true} \mid \text{ scale parameter c}) \quad (4)$$

Although not a true Gibbs sampler, experimental evidence indicates that the pseudo-Gibbs sampler generates reproducible estimates of these marginals from random initial configurations – an empirical proof of convergence. Note that the dependence of the posterior probabilities on the scale parameter c is explicitly recognized in the conditional probability notation.

A second user-provided parameter is the probability threshold γ. An appropriate choice of γ is determined empirically once the sampler has run its course. Histograms of marginal posterior probabilities, such as the one below, suggest that most markers are either isolated (p.p. near 0) or part of a larger block (p.p. near one).

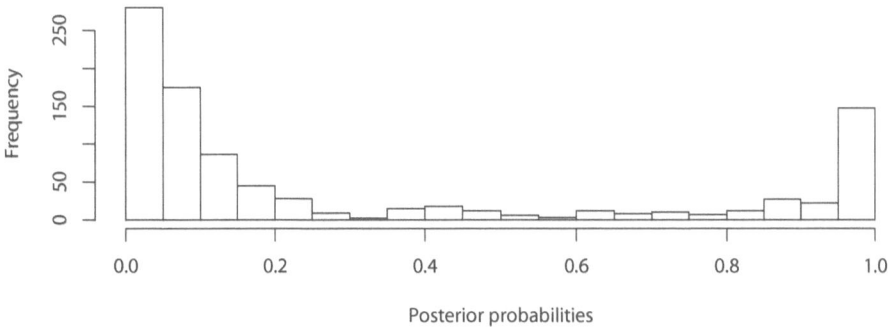

Fig. 1. Histogram of posterior probabilities using the medium resolution settings in Table 1 for a set of 938 gene-based markers present in four species. Discrimination is typically more pronounced with nucleotide-based markers.

The relative frequencies in (4) induce a stochastic ordering on $\mathcal{M}$. The stochastic ordering ranks markers by the strength of the evidence that each m joins with its neighbors to form a conserved segment. Recall that every configuration partitions $\mathcal{M}$ into two disjoint subsets $\mathcal{M}^{in}$ and its complement. Ranking $\{\pi_i(c)\}$ forms a family of partitions $\mathcal{M}^{in}(\gamma; c)$. Partitions of interest generally involve thresholds between 0.25 and 0.75. Hence, only a few configurations need actually be examined by the scientist.

Although ranking markers by their p.p. is a fairly robust procedure with respect to the scale parameter c, it is far from invariant. Large values of c omit small collinear blocks in favor of long blocks over large spans. Such runs are called low resolution. Conversely, high resolution (small c) runs identify small blocks that can disrupt larger low resolution blocks.

4   Results

Streptococcal strains are responsible for a wide range of diseases in humans. *S. pyogenes*, most commonly associated with "strep throat", also causes pneumonia or rheumatic fever if untreated [19]. *S. agalactiae* is the leading cause of pneumonia and meningitis in newborns [20]. *S. pneumoniae* [21] also causes of pneumonia and meningitis, but has multiple phenotypes that distinguish it from the other two species. Finally, *S. mutans* [22] is responsible for a large percentage of tooth decay. More Streptococcal genomes (nine) have been sequenced than any other genus. Although detailed comparative analyses have been conducted within species [20, 23] little has been published about all four species beyond pairwise contrasts (see [24] Table 3, [6] Supplemental material, and [20] Figure 2). Curiously, these previous analyses show that the smallest genome, *S. pyogenes*, and the largest genome, *S. agalactiae*, are the two most closely related taxa. By contrast, *S. pneumoniae* is the most phylogenetically distant species.

We present an analysis of genome rearrangements among these diverse strains using our pseudo-Gibbs sampler. To generate a set of monotypic markers, a reciprocal best BLAST search was done between each pair of Streptococcal genomes, retaining only matches with an E-value <0.00005 covering at least 50% of both proteins. The distribution of common reciprocal best matches among the four taxa are shown in Figure 2. Although 968 genes common to the four genomes meet these criteria in all six paired comparisons, only 938 consist exclusively of one-to-one matches within each comparison. This reduced group of genes, from which putative gene duplications have been removed, forms a set of monotypic markers. Given our stringent match criteria, most scientists would categorize these genes as orthologs. We use these markers to investigate two complementary aspects of comparative genomics: identifying connected neighborhoods of orthologous genes and inferring ancestral genome architecture. These two problems demand different levels of resolution to identify rearranged segments.

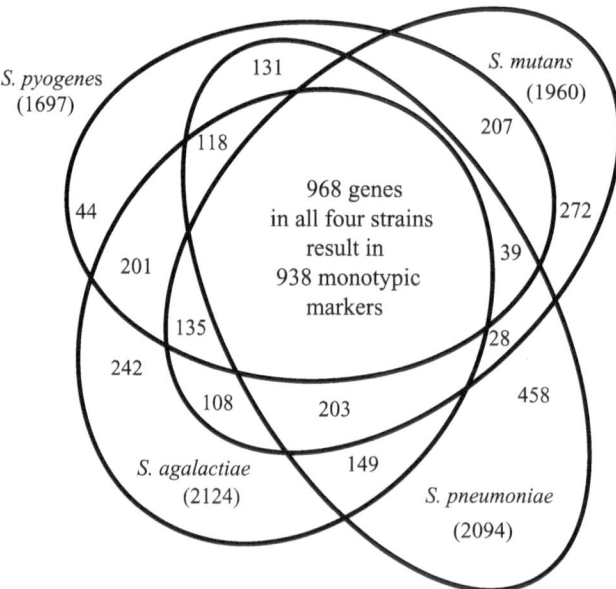

Fig. 2. This Venn diagram shows the partition of genes from all four Streptococcal genomes into 15 groups of mutual reciprocal best BLAST hits. *S. pneumoniae* has the most lineage-specific genes (458), while *S. pyogenes* has only 44 unique genes. Removing paralogous genes leaves 938 monotypic markers common to all four species.

The scale parameter c divides the score (1), affecting the size of detectable gene clusters. A smaller value of c generally improves sensitivity to small collinear segments but may introduce additional noise. The threshold parameter γ partitions markers into signal and noise: markers with p.p.$> \gamma$ are deemed signal while the rest are considered noise. As mentioned above, γ is can be chosen by inspecting the frequency distribution of posterior probabilities (see Figure 1). Breakpoints in the selected set of signal markers define collinear segments of the genomes under study.

We begin with a search for clusters of orthologous genes. Since the introduction of clusters of orthologous groups, or COGs [25, 26], the concept has been expanded to include connected gene neighborhoods [27, 28]. Typically methods to construct gene neighborhoods start with triplets of genes, and inductively work their way up to larger connected neighborhoods. Rather than growing a neighborhood from a 'seed' COG, our method directly identifies neighborhoods as locally collinear blocks among the genomes. Statistical tests have been developed for assessing patterns of collinearity between two genomes [30, 29], but they have not been extended to multi-genome analysis.

Our empirical approach permits us to indirectly modulate the minimum neighborhood size by adjusting γ and c. Table 1 shows a series of runs on the 938 Streptococcus markers using different parameter settings to achieve low, medium, and two high resolutions. In particular, observe that both 17-gene clusters are split into smaller LCBs at high resolution.

Table 1. Distribution of gene counts per collinear segment. Clusters are determined under three different conditions ranging in resolution from high to low. As the resolution increases, some large clusters split into smaller clusters by the emergence of a previously unnoticed intervening cluster.

Number of Segments (LCBs)

Genes per segment	2	3	4	5	6	7	8	9	10	11	13	14	17	18	24	Total
Resolution parameters (c, γ, w_{min})																
Low (75,45%,20)			1	2	1	6	2	5	3	1			2	1	2	26
Medium (30,45%,8)		3	5	6	2	6	1	4	1	1			2	1	2	34
High-1 (20,50%,15)	1	4	20	7	7	2	6	2	3	1	2			1	2	57
High-2 (20,30%,15)	3	11	29	7	7	2	6	1	3	1	2	1			2	72

The four runs in Table 1 use a score function where the current marker's weight is reduced 25 %. The role of λ in (1) is apparent when the medium resolution run in Table 1 is repeated without it (*i.e.* $\lambda = 1$) and the runs are compared. The number of LCBs jumps from 36 to 109, including 20 singletons and 15 pairs. Several large blocks are split into smaller segments, contributing to the increase while obscuring the underlying pattern of collinearity.

The same phenomenon can occur if γ is lowered. We present a particularly interesting example in Figure 3, magnified so genes can be represented as rectangles of varying length rather than points.

The large black collinear region in Figure 3 merits special attention. The smallest gene contributing to this segment is the ribosomal protein L34. At the lower threshold, L34 becomes isolated by the group of genes immediately to its right in *S. pneumoniae*, labeled "four gene cluster", and the dltDCAB operon to its left in *S. agalactiae*. Note that the largest genome, *S. agalactiae*, has the fewest number of lineage-specific genes within this segment.

We applied GRIMM [34] and MGR [35] to infer the ancestral genome organization of the four stains. Our initial analysis used high resolution collinear

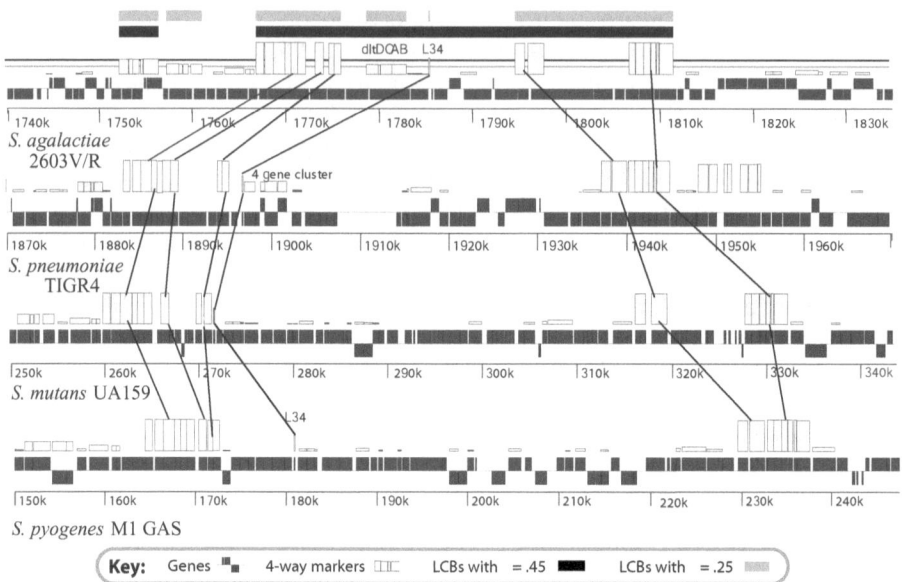

Fig. 3. A 100 kb region of *S. agalactiae* and the corresponding regions in the other three Streptococcus genomes. Meaning of rows within genome panel, starting at the bottom: genome coordinates, annotated genes (in black) with vertical position denoting transcriptional direction. Monotypic gene markers are drawn as white boxes with heights proportional to their posterior probability. Lines among the genomes connect clusters of markers that together comprise a large LCB. Other markers are outside the field of view in at least one genome and thus connecting lines can't be drawn. Black bars across the top denote LCBs formed from the medium resolution settings in Table 1. Lowering γ to 0.25 interjects small segments that break up the large black block.

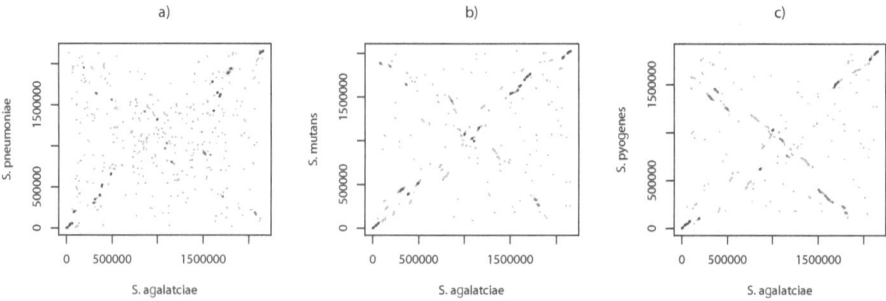

Fig. 4. X-plots of the 938 common orthologs, with each gene's position in *S. agalactiae* on the horizontal axis plotted against the corresponding position in each of the three other strains. Many putative orthologs do not meet the p.p. threshold at low resolution, and are drawn in grey. Genes above the threshold are black. In (c), and to a lesser degree in (b), collinear segments near the center of the X-plot are visible in grey. This suggests that certain orthologs may be collinear in some genomes, but not in all four.

segments from Table 1. The rearrangement scenarios suggested by MGR on high resolution segments do not maintain replichore balance, indicating that some of the collinear segments in this data set may not have been rearranged by symmetric inversions (data not shown). We then examined the 26 segments at from the low resolution run. Markers that exceed the p.p. threshold. tend to cluster about the diagonals of the X-plots shown in Figure 4, a pattern consistent with multiple symmetric inversions.

Using the 26 low-resolution segments, we ran GRIMM and MGR again. The result is a collection of 37 inversion events that maintain replichore balance among genomes (data not shown). Figure 5 shows a phylogenetic tree based on inversion events with branch labels giving the number of inversions per branch. Unlike some other comparisons [36], the frequency of genomic rearrangements between *Streptococci* appears to correlate well with the overall level of sequence divergence.

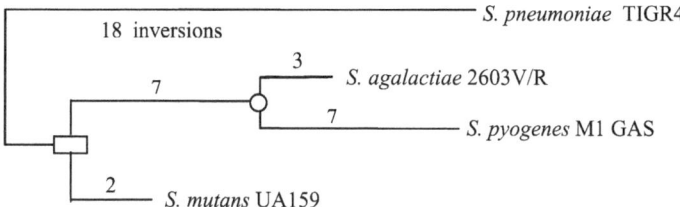

Fig. 5. Phylogeny of Streptococcal strains based on a parsimonious set of 26 genomic rearrangements of large segments, courtesy of MGR and GRIMM. The number of inversions along each lineage accompanies the branch. The circle denotes the ancestral genome of *S. pyogenes* and *S. agalactiae*; the rectangle is the ancestral genome of the circle and *S. mutans*.

The comparison among these four Streptococci allows us to infer the ancestral organization of the MRCA of *S. agalactiae*, *S. pyogenes.*, and *S. mutans*, by using *S. pneumoniae* as an outgroup. A separate 3-way analysis could be conducted for the MRCA of *S. agalactiae* and *S. pyogenes* with *S. mutans* as the outgroup.

5   Discussion and Conclusion

By assigning posterior probabilities to monotypic markers our algorithm assists in making a "best guess" as to which LCBs constitute evolutionarily conserved segments among a group of genomes. Once identified, such conserved segments can be subject to further analyses such as multiple global alignment or phylogenetic inference of genome organization. Furthermore, the flexibility of our algorithm makes it well suited to comparisons of both eukaryotic and prokaryotic genomes. By using markers based on inexact protein or nucleotide sequence matches the algorithm accommodates significantly diverged genomes, and its ignorance of distance between markers allows it to be applied to genomes with significant lineage-specific content.

In eubacteria, the origin and terminus of replication divide a circular chromosome into two replichores of similar length. Equal sized replichores are thought to maximize efficiency of replication of the genome and an imbalance of more than 20% can be selected against [31]. It is currently believed that symmetric inversions are the predominant means of genome rearrangements in eubacteria [5, 6, 32]. GRIMM and MGR implement sorting by reversals for circular chromosomes without any constraint on the replichore sizes of ancestral intermediates. As such, these tools are not appropriate when small clusters of orthologous genes are included – if in fact they were translocated by some other means [33]. We stress that it is the scientist's responsibility to judge whether such tools provide a useful analysis of a given bacterial data set.

If inversions are not responsible for all such "micro-rearrangements", other evolutionary mechanisms must be. One explanation for the observed micro-rearrangements is transposition mediated by insertion sequences. An alternative explanation is parallel lateral gene transfer events, acting independently to introduce the same DNA to different loci in each lineage. This phenomenon is called convergent evolution. A related mechanism is serial evolution - a horizontal transfer of DNA into one lineage followed by a transfer from that lineage to a second one. A fourth possibility would be ancient gene duplication and subsequent loss of the original gene copy. Attributing the mechanism responsible for a particular micro-rearrangement remains an open problem.

Acknowledgements

We thank Guillaume Bourque and Glenn Tesler for help with MGR and GRIMM, and Elisabeth Tillier for a critical reading of a related manuscript. Funding for all authors was provided by NIH Grant GM62994-02. A.E.D. was supported in part by NLM Training Grant 5T15M007359-03.

References

1. Nadeau, J.H., Taylor, B.A.: Lengths of chromosomal segments conserved since divergence of man and mouse. *Proc Natl Acad Sci USA* **81** (1984) 814-818
2. Waterston RH, Lindblad-Toh K, Birney E, Rogers J, Abril JF, Agarwal P, Agarwala R, Ainscough R, Alexandersson M, An P et al: Initial sequencing and comparative analysis of the mouse genome. *Nature* **420** (2002) 520-562.
3. Pevzner P, Tesler G: Genome rearrangements in Mammalian evolution: lessons from human and mouse genomes. *Genome Res* 2 **13**(2003) 37-45.
4. Schmid MB, Roth JR: Selection and endpoint distribution of bacterial inversion mutations. *Genetics* **105**(1983) 539-557.
5. Eisen JA, Heidelberg JF, White O, Salzberg SL: Evidence of symmetric chromosomal inversions around the replication origin in bacteria. *Genome Biology* **1**(2000) 1-9.

6. Tillier ER, Collins RA: Genome rearrangement by replication-directed translocation. *Nat Genet* **26**(2000) 195-197.
7. Blanchette M, Kunisawa T, Sankoff D: Gene order breakpoint evidence in animal mitochondrial phylogeny. *J Mol Evol* **49**(1999) 193-203.
8. Fitch WM: Homology a personal view on some of the problems. *Trends Genet* **16**(2000) 227-231.
9. Darling ACE, Mau B, Blattner FR, Perna NT: Mauve: Multiple Alignment of Conserved Genomic Sequence with Rearrangements. *Genome Res* **14** (2004) 1394-1403.
10. Calabrese PP, Chakravarty S, Vision TJ: Fast identification and statistical evaluation of segmental homologies in comparative maps. *Bioinformatics* **19** (2003) Suppl 74-80.
11. Darling A, Mau B, Blattner FR, Perna NT: GRIL: Genome rearrangement and inversion locator. *Bioinformatics* **20** (2003) (122-124).
12. Buhler J: Efficient large-scale sequence comparison by locality-sensitive hashing. *Bioinformatics* **17** (2001) 419-428.
13. Ma, B, Tromp J, Li M: PatternHunter: faster and more sensitive homology search. *Bioinformatics* **18** (2002) 440-445.
14. Brudno M, Steinkamp R, Morgenstern B: The CHAOS/DIALIGN WWW server for multiple alignment of genomic sequences. *Nucleic Acids Res.* (Web Server issue) (2004) W41-44.
15. Schwartz S, Kent WJ, Smit A, Zhang Z, Baertsch R, Hardison RC, Haussler D, Miller W: Human-mouse alignments with BLASTZ. *Genome Res* **13**(2003) 103-107.
16. Altschul SF, Madden TL, Schaffer AA, Zhang J, Zhang Z, Miller W, Lipman DJ: Gapped BLAST and PSI-BLAST: a new generation of protein database search programs. *Nucleic Acids Res* **25** (1997) 3389-3402.
17. Besag J: Spatial Interaction and the Statistical Analysis of Lattice Systems. *Journal of the Royal Statistical Society Series B* **36** (1974) 192-236.
18. Tierney L: Markov chains for exploring posterior distributions. *Annals of Statistics* **22** (1994) 1701-1762.
19. Ferretti JJ, McShan WM, Ajdic D, Savic DJ, Savic G, Lyon K, Primeaux C, Sezate S, Suvorov AN, Kenton S et al: Complete genome sequence of an M1 strain of Streptococcus pyogenes. *Proc Natl Acad Sci U S A* **98** (2001) 4658-4663.
20. Tettelin H, Masignani V, Cieslewicz MJ, Eisen JA, Peterson S, Wessels MR, Paulsen IT, Nelson KE, Margarit I, Read TD et al: Complete genome sequence and comparative genomic analysis of an emerging human pathogen, serotype V Streptococcus agalactiae. *Proc Natl Acad Sci U S A* **99** (2002) 12391-12396.
21. Tettelin H, Nelson KE, Paulsen IT, Eisen JA, Read TD, Peterson S, Heidelberg J, DeBoy RT, Haft DH, Dodson RJ et al: Complete genome sequence of a virulent isolate of Streptococcus pneumoniae. *Science* **293** (2001) 498-506.
22. Ajdic D, McShan WM, McLaughlin RE, Savic G, Chang J, Carson MB, Primeaux C, Tian R, Kenton S, Jia H et al: Genome sequence of Streptococcus mutans UA159, a cariogenic dental pathogen. *Proc Natl Acad Sci U S A* **99** (2002) 14434-14439.
23. Smoot JC, Barbian KD, Van Gompel JJ, Smoot LM, Chaussee MS, Sylva GL, Sturdevant DE, Ricklefs SM, Porcella SF, Parkins LD et al: Genome sequence and comparative microarray analysis of serotype M18 group A Streptococcus strains associated with acute rheumatic fever outbreaks. *Proc Natl Acad Sci U S A* **99** (2002) 4668-4673.

24. Ferretti JJ, Ajdic D, McShan WM: Comparative genomics of streptococcal species. *Indian J Med Res* **119** (2004) Suppl 1-6.
25. Tatusov RL, Koonin EV, Lipman DJ: A genomic perspective on protein families. *Science* 1997, **278** (1997) 631-637.
26. Tatusov RL, Natale DA, Garkavtsev IV, Tatusova TA, Shankavaram UT, Rao BS, Kiryutin B, Galperin MY, Fedorova ND, Koonin EV: The COG database: new developments in phylogenetic classification of proteins from complete genomes. *Nucleic Acids Res* **29** (2001) 22-28.
27. Rogozin IB, Makarova KS, Murvai J, Czabarka E, Wolf YI, Tatusov RL, Szekely LA, Koonin EV: Connected gene neighborhoods in prokaryotic genomes. *Nucleic Acids Res* **30** (2002) 2212-2223.
28. Omelchenko MV, Makarova KS, Wolf YI, Rogozin IB, Koonin EV: Evolution of mosaic operons by horizontal gene transfer and gene displacement in situ. *Genome Biol* **4** (2003) R55.
29. Hampson, S, McLysaght, A, Gaut, BS, Baldi, PF: LineUp: Statistical Detection of Chromosomal Homology with Application to Plant Comparative Genomics. *Genome Research* **13** (2003) 999-1010.
30. Durand, D, Sankoff, D: Tests for Gene Clustering. *Journal of Computational Biology.* **10** (2003) 453-482.
31. Guijo MI, Patte J, del Mar Campos M, Louarn JM, Rebollo JE: Localized Remodeling of the Escherichia coli Chromosome. The patchwork of segments refractory and tolerant to inversion near the replication terminus. *Genetics* **157** (2001) 1413-1423.
32. Ajana, Y., Lefebvre, J. F., Tillier, E., El-Mabrouk, N.: Exploring the set of all minimal sequences of reversals - An application to test the replication-directed reversal hypothesis. Second International Workshop, Algorithms in Bioinformatics (WABI 2002), *LNCS* **2452**, R. Guigo and D. Gusfield eds. (2002) 300-315.
33. Lefebvre JF, El-Mabrouk N, Tillier ER, Sankoff D: Detection and validation of single gene inversions. *Bioinformatics* **19,** Suppl. 1, special issue, 11th International Conference on Intelligent Systems for Molecular Biology (2003) 190-196.
34. Tesler G: GRIMM: genome rearrangements web server. *Bioinformatics* **18** (2002) 492-493.
35. Bourque G, Pevzner PA: Genome-scale evolution: reconstructing gene orders in the ancestral species. *Genome Res* **12** (2002) 26-36.
36. Deng W, Burland V, Plunkett G 3rd, Boutin A, Mayhew GF, Liss P, Perna NT, Rose DJ, Mau B, Zhou S, Schwartz DC, Fetherston JD, Lindler LE, Brubaker RR, Plano GV, Straley SC, McDonough KA, Nilles ML, Matson JS, Blattner FR, Perry R: Genome sequence of Yersinia pestis KIM. *J Bacteriol.* **184** (2002) 4601-4611.

Genome Rearrangement in Mitochondria and Its Computational Biology

István Miklós[1] and Jotun Hein[2]

[1] Hungarian Academy of Science and Eötvös Loránd University of Science,
Theoretical Biology and Ecology Group
1117 Budapest, Pázmány Péter sétány 1/c, Hungary
miklosi@ramet.elte.hu
[2] Oxford Centre for Gene Function
University of Oxford
South Parks Road, Oxford, OX1 3QB, UK
hein@stats.ox.ac.uk

Abstract. In the first part of this paper, we investigate gene orders of closely related mitochondrial genomes for studying the properties of mutations rearranging genes in mitochondria. Our conclusions are that the evolution of mitochondrial genomes is more complicated than it is considered in recent methods, and stochastic modelling is necessary for its deeper understanding and more accurate inferring. The second part is a review on the Markov chain Monte Carlo approaches for the stochastic modelling of genome rearrangement, which seem to be the only computationally tractable way to this problem. We introduce the concept of partial importance sampling, which yields a class of Markov chains being efficient both in terms of mixing and computational time. We also give a list of open algorithmic problems whose solution might help improve the efficiency of partial importance samplers.

1 Introduction

The idea that differences between the gene orders of two genomes can be used as a measurement of evolutionary distance was proposed more than six decades ago [1]. It was rediscovered in the eighties [2], and since then a large set of papers on optimisation methods for genome rearrangement problems has been published. However, except the case of sorting signed permutations by inversions [3–7] or by translocations [8], only approximations [9–13] and heuristics [14] exist. Most of the methods concerning with more types of mutations either penalise all the mutations with the same weight [12], or exclude a whole set of possible mutations due to a special choice of weights [11].

The principle of choosing solutions by minimising the amount of evolution is also called parsimony and has been widespread in phylogenetic analysis. Over the last two decades the parsimony method of phylogenetic reconstruction has been severely criticised and lost terrain to methods based on stochastic modelling of evolution [15, 16]. Statistical methods give not only more consistent estimations, but also the possibility of hypotheses testing and goodness-of-fit testing

J. Lagergren (Ed.): RECOMB 2004 Ws on Comparative Genomics, LNBI 3388, pp. 85–96, 2005.
© Springer-Verlag Berlin Heidelberg 2005

of the underlying model. Therefore it is a natural attempt to develop statistical methods also for genome rearrangement.

Recently a few papers were published on probability models of genome rearrangement. Two of them considered only inversions [17, 18], and a third one worked on multi-chromosomal genomes rearranged by both inversions and translocations [19]. We also started a Bayesian approach to the genome rearrangement problem considering insertions, transpositions and inverted transpositions in unichromosomal genomes [20, 21]. As we will explain it in detail in Section 4, all of these methods have the same computational concept: instead of analytical solutions, they apply a Markov chain Monte Carlo (MCMC) converging to the distribution defined by the underlying stochastic model. Samples from this Markov chain estimate statistics of interest like the posterior distribution of number of mutations happened, evolutionary parameters, etc.

All the underlying models used so far in genome rearrangement problems are relatively simple, and an obvious way to achieve further improvements on this methodological line is to improve models describing the evolution of gene orders. We will focus on Metazoa mitochondrial genomes for several reasons. Mitochondrial genomes are unichromosomal circular genomes consisting of usually 13 protein coding genes and 24 RNA genes. This gene content as well as the genes themselves are very conservative. Each gene is represented in one copy except a very few cases that helps to identify homologous gene pairs. Mitochondrial genomes are very compact lacking introns and transposons, which are known to influence the rate of mutations rearranging genomes. Previous results showed that there were no selection on any particular gene order in mitochondria.

The large amount of available mitochondrial genome data allows the investigation of closely related genomes. Our aim is to reveal elementary steps of the evolution of gene orders. In Section 3, we report our findings based on the investigation of the NCBI database on gene orders in mitochondria, and we review previous hypotheses, too. In Section 4, we describe a general framework for modelling the dynamics behind the evolution of gene orders in mitochondria. Efficient computation in biologically more realistic models needs a core logic being significantly different from that used in optimisation methods. We give a list of open algorithmic questions related to this framework to demonstrate that Markov chain Monte Carlo is not a boring technique but a source of interesting computational problems. In Section 5, we discuss how similar research could be conducted on bacterial or nuclear genomes.

2 Mathematical Description of Genome Rearrangement

We consider that two genomes have the same gene content, each gene is represented in one copy in both genomes. We describe their gene orders as signed permutations, numbers correspond to genes, signs represent the reading direction of genes. Based on elementary group theory, a series of mutations transforming a genome π_1 to genome π_2 also transforms a permutation $\pi_2^{-1}\pi_1$ to the identical permutation. Therefore we will always consider *sorting* permutations, namely transforming a permutation into the identical one. For circular genomes, we choose an anchor gene being in position 1 and having a positive sign. On the NCBI database, this is the NADH dehydrogenase subunit 1. It is easy to show that the rest of the genome mimics the evolution of a linear genome, hence the genome rearrangement problem of circular genomes having n genes is equivalent with the genome rearrangement problem of linear genomes with $n-1$ genes. We will talk about linear signed permutations in the rest of the paper. We follow the convention representing a signed permutation of length n as an unsigned permutation of length $2n$, we replace $+i$ with $2i-1, 2i$, and $-i$ with $2i, 2i-1$. This unsigned permutation is then framed to 0 and $2i+1$. Only segments $[2i+1, 2j]$ are allowed to mutate in the unsigned representation.

Starting with 0, we connect every other position in the permutation with a straight line, and starting also with 0, we connect every other number of the permutation with an arc. We consider the permutation as a graph, called *graph of reality and desire*, whose vertexes are the numbers from 0 to $2n+1$, and edges are the straight lines and arcs. The permutation can be unequivocally decomposed into cycles. Following the convention, we call the straight lines black edges or reality edges, and arcs are named grey edges or desire edges. A black edge is a breakpoint if its cycle is longer than a black edge and a grey arc. The breakpoint distance between genomes π_1 and π_2 is the number of breakpoints in $\pi_2^{-1}\pi_1$.

3   Inferring Closely Related Mitochondrial Genomes

To date, 466 fully sequenced Metazoa mitochondrial genomes have been published in the NCBI database yielding 140 different gene orders. For these 140 genomes, all pairs of genomes were compared. Compared pairs were sorted based on their breakpoint distance. Results can be downloaded form the World Wide Web, http://www.stats.ox.ac.uk/~miklos/metazoa.tar.gz.

Among the $\binom{140}{2} = 9730$ comparisons, 36 have 0 breakpoint distance. Although these pairs have different gene content, the intersection of genes have the same order. All of the genomes in these comparisons belong to the Vertebrata phylum.

A breakpoint distance 1 is mathematically impossible. We found 72 genome pairs having breakpoint distance 2. Remarkably, 70 of them are caused by inversions of a single gene. In all of these cases, the inverted gene was a tRNA gene. One of the long insertions is two long (*Pollicipes polymerus* vs. *Tetraclita japonica*), and the second one is eighteen long (*Pisaster ochraceus* vs. several urchins having the same gene order).

Breakpoint distance 3 is always caused by a transposition, an inverted transposition or two adjacent inversions. We found 146 comparisons with breakpoint distance 3. Several of these pairs differ in a swap of two adjacent genes. A frequent pattern is a transposition of a single gene that is moved far from its original place. There are several cases for inverted transpositions, even a two long fragment can be inverted, for example *Caelorinchus kishinouyei* vs. *Gonostoma gracile*. Remarkably again, we did not found any transposed fragment longer than two. Two adjacent inversions acting on neighbour tRNA genes coding cystein and tyrosine might also cause a breakpoint distance 3 in several vertebrates.

The first really interesting breakpoint distance is 4, because it cannot be caused by a single inversion, transposition or inverted transposition. In terms of such mutations it might be the result of two non-adjacent inversions or two mutations having a common black edge, of which at least one of them is a transposition or inverted transposition. Surprisingly, behind almost half of the breakpoint distances 4 we can see the later configuration (40 from 93). In most of these surprising cases, the pattern can be described with a transposition followed by the inversion of a single gene at one of the ends of the inverted fragment. If mutations happened independently we would see about one tenth of the inversion-transposition pairs connected (recall that the vast majority of inversions acts on a single gene). We can say that connected inversion-transposition pairs are more frequent than we would expect considering independence, 35 connected pairs versus 121 independent pairs having breakpoint distance 5. However, it is hard to elaborate statistical significance, since genomes were not sequenced randomly but based on biological interest, hence we have a biased sample. In all cases, the common black edge was in the vicinity of a control region. It has already been reported that genome rearrangements are more frequent around the control region [22].

We show two examples on Fig. 1. A transposition and an inversion separates the *O. bicirrhosum* mitochondrial genome from several bird genomes. Next to the common point of the two mutations lies the control region in both genomes. Fig. 1b shows genome rearrangements between ticks *I. hexagonus* and *R. sanguineus*. In *R. sanguineus*, a new control region has emerged. This new CR clearly attracts mutations.

We must mention that there might be mutations increasing significantly the breakpoint distance. Boore proposed a duplication-loss model a few years ago [22]. In this model, a part of the genome is duplicated, and after the duplication, one copy of all gene pairs are eliminated. Boore found an excellent example strongly supporting this model. In primitive Holothuroidea, a set of 15 tRNA genes can be found in one part of the mitochondrial genome. In *Cucumaria*, 9 of them stay in the same place, while six of them moved to another part of the genome. Only a couple of these six were adjacent in the original rearrangement, however, all of them kept the reading direction and their relative order. Moreover, several nonassignable nucleotides can be found in *Cucumaria* exactly at those places where unused genes are being eliminated.

We might conclude that all the three classical mutation types (inversions, transpositions and inverted transpositions) occur during the evolution of mitochondrial gene orders, and these mutations usually act on short segments and they are frequently correlated. The correlation is suspiciously caused by control regions. Other type of mutations should be also considered. A duplication-loss model was already published, a type of mutation that can cause significant increase of breakpoint distance in one step. Since there is a selection pressure keeping mitochondrial genomes very compact, the speed of gene loss should be

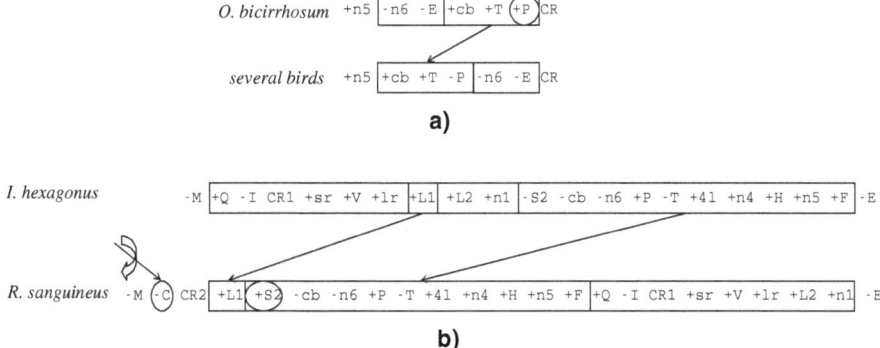

Fig. 1. Genome rearrangement between **a)** several birds and *Osteoglossum bicirrhosum* and **b)** ticks *Ixodes hexagonus* and *Rhipicephalus sanguineus*. Arrows show transpositions, circles show inversions of single genes, arrow with a twisted arrow on it indicates inverted transposition. This transposed and inverted tRNA-Cys gene originated from a region not shown in the figure. Abbreviations: **n1, n4, n5**: NADH dehydrogenase subunit 1, 4 and 5; **sr**: 12S rRNA; **lr**: 16S rRNA; **CR, CR1, CR2**: control regions; tRNA genes are denoted by the one-letter amino acid abbreviation.

very fast, and the entire scenario can be modelled as a single mutation happening in infinitesimally small time.

All these findings might question the usefulness of methods considering only inversions. In the next section, we give a review on more complicated models for genome rearrangement and their computational properties. We do not have definite solutions on how to treat duplication-loss events, but we describe an approach which might be fruitful for correlated mutations.

4   Stochastic Modelling of Genome Rearrangements

4.1   Time-Continuous Markov Model

Time-continuous Markov models have been the standard approach for stochastic modelling of molecular evolution. In case of genome rearrangements, points in the state space of the Markov model are the possible gene orders. The number of these points grows with the factorial of the number of genes, therefore the brute force calculation working fine for example for substitution models of nucleic acids is computationally intractable here. What we can calculate is the likelihood of a trajectory, which is the probability that a given sequence of mutations happened in a time span conditional on a set of parameters describing the model. It is easy to find a closed form for this likelihood when the sum of the rates of mutations does not depend on the actual state [20, 23], it is

$$e^{-tr \times t} \frac{\prod_{i=1}^{l} r_i t}{l!} \tag{1}$$

where tr is the sum of rates of all possible mutations at each state (total mutation rate or *exit rate*, see [23]) t is the spanning time of the trajectory, r_i is the rate of the ith mutation, and l is the length of the trajectory. The exit rate always keeps constant when we work on unichromosomal genome, insertions and deletions are excluded and mutations happen independently. Durrett *et al.* recognised that the sum of rates changes along the trajectory for multi-chromosomal genomes, and they introduced pseudo-events to maintain the constant total rate along the trajectory [19]. When the rate of mutations depends on the positions of previous mutations, the sums of rates are changing along the trajectory, since different mutations act on different number of positions, and hence in biologically more realistic problems, we will face with the same problem, even for unichromosomal genomes. Duplications and deletions or emergence of a new control region also cause the exit rate changing. The approximation introducing pseudo-events is not necessary, since exact likelihood values can be calculated even when the sum of the rates are varying with states of the Markov model. However, exact calculations need more computational time, the state-of-the-art algorithm [23] calculating trajectory likelihoods runs in $O(l^2)$ time, where l is the length of the trajectory, namely, the number of states in it.

4.2 Metropolised Partial Importance Sampler: A New MCMC Strategy

The aim is to sample trajectories with probabilities proportional to their likelihood, and this can be done by using Markov chain Monte Carlo (MCMC) [24, 25]. In all of the published methods, an update in the Markov chain replaces a part of the trajectory. The proposal for the new sub-trajectory is drawn from a distribution that mimics the target distribution we would like to sample from, and the new proposal is independent from the old sub-trajectory. We call this strategy *Metropolised Partial Importance Sampler* [21]. The success of such a sampler is based on how well we can approximate a target distribution in small dimensions. Here the target distribution is the posterior distribution of trajectories given two genomes that the trajectories should connect. If genomes are close to each other, trajectories are short (low dimension). When we resample short sub-trajectories, the two rearrangements at the ends of the sub-trajectory are similar.

The discrepancy between the proposal and the target distribution is corrected by accepting the proposal with probability

$$\min\left\{1\ ,\ \frac{P(X|Y)\pi(Y)}{P(Y|X)\pi(X)}\right\} \tag{2}$$

where P is the proposal distribution, π is the target one, X is the actual state of the chain, and Y is the proposal, and the chain remains in state X with the complement probability. The probability in Equation 2 is known as the Metropolis-Hastings ratio [24, 26]. The beauty of MCMC is that whenever the Markov chain based on the proposals is ergodic on the state space of the target distribution, and

$$P(X|Y) > 0 \quad \text{if and only if} \quad P(Y|X) > 0 \tag{3}$$

the Metropolis-Hastings ratio transforms it into a chain converging to the target distribution. However, the convergence might be very slow, and a slow convergence also implies high correlation between neighbour points of the chain. From a computational point of view it is also important question how much time it takes to get a proposal and to calculate the Metropolis-Hastings ratio. We are going to discuss both mixing and computational properties of Markov chain Monte Carlo for genome rearrangement.

Mixing of Markov Chains. For fast convergence and good mixing, the proposal distribution should be selected carefully. An ideal proposal distribution has low dependence on the actual state and yields Metropolis-Hastings ratios always close to 1. Indeed, if the proposal did not depend on the actual state of the chain and the Metropolis-Hastings ratio was always 1, namely, if

$$P(Y|X) = P(Y) = \pi(Y) \tag{4}$$

then the Markov chain would consist of independent samples of the target distribution. (Although a sampler might be super-efficient having smaller sampling variance than that of independent sampling [25], we will not discuss it here.) The proposed sub-trajectory does not depend on the removed one, therefore the dependency of the new trajectory correlates only with the length of the sub-trajectory being resampled. From this point of view, resampling long sub-trajectories seems to be a good idea. On the other hand, we can sample from a distribution only *approximating* the target one. The overlapping of the proposal and target distributions will obviously decrease with the length of the resampled sub-trajectory causing small Metropolis-Hastings ratio, and hence, small acceptance ratio. An additional observation is that the proposed and original sub-trajectories might have different length, therefore all possible length for sub-trajectories should be proposed to satisfy Condition 3. We conjecture that it is very hard to prove that a particular type of distribution is better than another for the purpose. However, several types of distributions were used successfully in the literature [18, 20]. Due to the trade-off between dependency in the Markov chain and acceptance ratio, there should be an optimal expected length of the resampled sub-trajectory for which the acceptance ratio is still sufficiently high, while on the other hand, the autocorrelation of the samples got small. The distribution can be optimised in preliminary performance studies [18].

The mixing of the Markov chain also depends on how well the proposal distribution can mimic the target distribution. Published methods propose new paths step by step. They measure the departure of the actual rearrangement from the rearrangement which the sub-trajectory must arrive to, and propose a mutation which decrease the measurement of the departure ('good' mutations) with high probability and propose other ones ('bad' mutations) with low probability. This philosophy seems to be essential, since random mutations would reach the target rearrangement with a very small probability. Based on the measurement of goodness of mutations, there are several possibilities to propose good and bad mutations differing also in the amount of computational time needed. We are going to discuss them below.

Efficiency of Sampling Sub-trajectories. The inversion distance problem is one of the few problems in genome rearrangement which has polynomial time solution and has been most intensively investigated [3, 4]. It is also possible to enumerate all sorting inversions, namely, inversions decreasing the inversion distance [7]. Similarly we can enumerate inversions that do not change or increase the inversion distance. An obvious strategy uses these enumerations for sampling sub-trajectories [17]: it samples a sorting inversion with high probability (uniformly among them), and non-sorting inversion with small probability. However, the running time for sampling a mutation grows with $\Omega(n^2)$, where n is the number of genes, hence it gets inefficient for large genomes.

An alternative strategy distinguishes inversions based on how they change the number of cycles in the graph of desire and reality [18, 20]. We distinguish three groups of inversions, +1-, 0- and −1-inversions, based on whether they change the number of cycles by 1, 0 or −1, respectively, see Fig. 2. A sampling strategy samples uniformly inside the three group, but the group of +1-inversions with high probability, and with decreasing probability the groups of 0- and −1-inversions. Although we know that an inversion increasing the number of cycles is not necessarily a sorting inversion, it is still a good measurement, since the identical permutation has $n + 1$ cycles, where n is the number of genes, and all other rearrangements have less cycles. The advantage of this strategy is that we can calculate the number of inversions in each group in linear time:

$$\#\text{1-inversion} = \sum_{i=0}^{k} (l(c_i) - p(c_i)) \, p(c_i) \tag{5}$$

$$\#\text{0-inversion} = \sum_{i=0}^{k} \binom{l(c_i) - p(c_i)}{2} + \binom{p(c_i)}{2} \tag{6}$$

$$\# - \text{1-inversion} = \binom{n+1}{2} - $$
$$- (\#\text{1-inversion}) - (\#\text{0-inversion}) \tag{7}$$

where k is the number of cycles, $l(c_i)$ is the length of cycle c_i, and $p(c_i)$ is the number of positive black edges of cycle c_i, that is the number of black edges one passes left to right on a tour of the cycle [20].

In an earlier work [20], we sampled inversions with the rejection method [27, 25], namely, we sampled an inversion uniformly, and if it was an inversion of the prescribed type, we accepted it, otherwise we rejected and drew a new sample till success. It is also possible to sample inversions of a prescribed type in a running time growing linearly with the number of genes [21], which turns out to be more efficient in practise. For example, we sample +1-inversions in the following way. We create a list of positive and negative edges for each cycle. We first sample

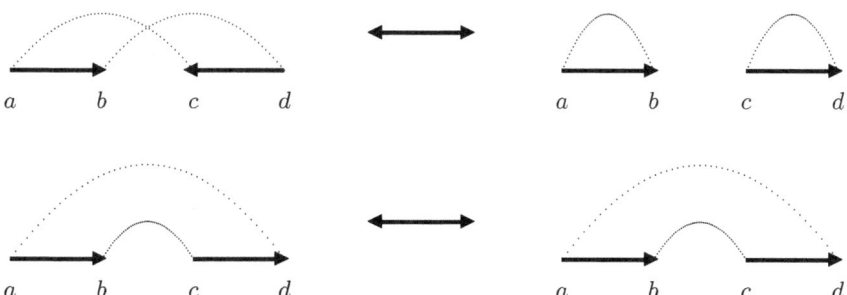

Fig. 2. How an inversion changes the number of cycles in a permutation. Note that each dashed line represents a path connecting two ends of black lines, which is not necessarily a single arc. A +1-inversion acts on two black edges of a cycle having different orientation. A −1-inversion acts on two cycles. A 0-inversion acts on two edges of a cycle having the same orientation.

a random cycle with a probability proportional to the number of +1-inversions acting on it, and then we choose a random positive and a random negative edge of this cycle uniformly. These two edges define the sampled +1-inversion. Due to the weighted sampling of cycles, the procedure guarantees uniform sampling on all +1-inversions. Similar strategies exist for uniform sampling of 0- and −1-inversions.

Sampling transpositions and inverted transpositions is not so well elaborated. Only one strategy is known, it distinguishes +2, +1 transpositions and inverted transpositions, and treats all the rest in the same way [20]. +2- and +1-mutations are listed, and sampled using this list, while other mutations are sampled with the rejection method. +2- and +1-transpositions and inverted transpositions act always on one cycle, and the enumeration investigates all the possible triplets of edges in all cycles. Although the running time for this sampler grows with the number of genes cubed in the worst case, in practise it is still fast, since long cycles are rare, especially when we resample short sub-trajectories. However, a better sampling strategy would be a great advantage.

It is not enough to somehow sample mutations from a reasonable distribution, but proposal and back-proposal probabilities ($P(Y|X)$ and $P(X|Y)$) should also be calculated. The rejection method is a good example that a sampling strategy is not necessarily able to calculate sampling probabilities. Sampling with rejection is always coupled with an algorithm enumerating the size of the set we are sampling from to get sampling probabilities. Therefore we need a thinking that might be unusual in optimisation strategies to improve present techniques in stochastic modelling: we need reasonable measurements for the goodness of a mutation which can be sampled fast, as well as we should be able to calculate proposal and back-proposal probabilities. In the next section we are going to list some open algorithmic problems.

4.3   Open Problems

Propose Transpositions in Sub-cubic Time. The state-of-the-art algorithm needs $O(n^3)$ time to propose a transposition from a reasonable distribution, where n is the number of genes in the genome. Can it be improved? Is there other distributions for which a faster strategy exist? Distinguishing transpositions based on the number of breakpoints they remove might be a promising way.

Polynomial Proposal for the Duplication-Loss Model. There are exponentially many possible duplication-loss events. How can we characterise good mutations and how can we efficiently sample from them?

Sub-squared Time Calculation for the General Trajectory Likelihood. Can we calculate trajectory likelihoods faster than $\Theta(n^2)$ time?

Reusing Information During Sub-trajectory Proposal. Can we update efficiently auxiliary variables storing information about cycle decomposition, number of different type of mutations, etc. during sampling trajectories using informations about the previous mutation sampled?

5   Discussion

Most of the methods in the scientific literature consider inversions, transpositions and inverted transpositions as elementary mutations rearranging genomes [28]. In this paper, we investigated mitochondrial genomes to get a better picture what elementary rearrangment events really are. We found all the tree types of mutations mentioned above, and we also showed that short rearrangements are frequent and mutations are not independent. The dependency is very likely caused due to the increased mutation rate around the control region.

In the second part of this paper, we gave an overview on recent progress in stochastic modelling genome rearrangement. Available techniques provide a partial solution how to incorporate more prior knowledge into the models to improve them. We can introduce different rates for different length of mutations as well as different rates for mutations acting around the control region without any problem, since they do not change the exit rate. Dependency on the acting points of the previous mutation does change the exit rate, since transpositions act on three black edges of the breakpoint graph, while inversions act only on two. An interesting problem would be to change samplings such that dependent mutations are proposed more frequently, but changing the likelihood calculations on its own enough to get a Markov chain converging to the desired new distribution. Modified likelihood calculations need increased running time, and this might be a drawback when trajectories are long.

Introducing dependency between mutations would be a simple model of more complicated mutations like duplication-loss events. Indeed, in such a model, a sequence of mutations mimicing a duplication-loss event would have higher likelihood than a sequence of similar mutations not having a common edge in the breakpoint graph. However, a reassuring solution would be the explicit modelling of these events.

The number of sequenced genomes grows quickly, and recent sequencing projects meet the requirement of comparative genetics to sequence closely related genomes. It would be worth investigating bacterial and Eukaryota nuclear genomes, as well. However, such a work will definitely be more complicated: there might be several copies of genes, hence we cannot describe a genome as a signed permutation. Additionally, pseudogenes, transposons, repetitive elements should be modelled in a reasonable way. In spite of the difficulties, the authors believe that stochastic modelling and MCMC will be the main key in modern comparative genetics.

Acknowledgements

This work was funded by EPSRC grant HAMJW and MRC grant HAMKA. I.M. was further funded by a Békésy György postdoctoral fellowship.

References

1. Sturtevant, A.H., Novitski, E.: The homologies of chromosome elements in the genus Drosophila. Genetics **26** (1941) 517–541
2. Palmer, J.D., Herbon, L.A.: Plant mitochondrial DNA evolves rapidly in structure, but slowly in sequence. J. Mol. Evol. **28** (1988) 87–97
3. Bader, D.A., Moret, B.M.E., Yan, M.: A linear-time algorithm for computing inversion distance between signed permutations with an experimental study. J. Comp. Biol. **8(5)** (2001) 483–491
4. Bergeron, A.: A very elementary presentation of the Hannenhalli-Pevzner theory. In: Proceedings of CPM2001 (2001) 106–117
5. Hannenhalli, S., Pevzner, P.A.: Transforming Cabbage into Turnip: Polynomial Algorithm for Sorting Signed Permutations by Reversals. Journal of ACM **46(1)** (1999) 1–27
6. Kaplan, H., Shamir, R., Tarjan, R.: A faster and simpler algorithm for sorting signed permutations by reversals. SIAM J. Comput. **29(3)** (1999) 880–892
7. Siepel, A.: An algorithm to find all sorting reversals. In: Proceedings of RE-COMB2002 (2002) 281–290
8. Hannenhalli, S.: Polynomial algorithm for computing translocation distance between genomes. In: Proceedings of CPM1996 (1996) 168–185
9. Bafna, V., Pevzner, A.: Sorting by transpositions. SIAM J. Disc. Math. **11(2)** (1998) 224–240
10. Berman, P., Hannenhalli, S., Karpinski, M.: 1.375-Approximation Algorithm for Sorting by Reversals. In: Proceedings of ESA2002 (2002) 200–210
11. Eriksen, N.: $(1+\varepsilon)$-approximation of sorting by reversals and transpositions. In: Proceedings of WABI2001, LNCS **2149** (2001) 227–237
12. Gu, Q-P., Peng, S., Sudborough, H.I.: A 2-Approximation Algorithm for Genome Rearrangements by Reversals and Transpositions. Theor. Comp. Sci. **210(2)** (1999) 327–339

13. Kececioglu, J.D., Sankoff, D.: Exact and Approximation Algorithms for Sorting by Reversals, with Application to Genome Rearrangement. Algorithmica **13(1/2)** (1995) 180–210
14. Blanchette, M., Kunisawa, T., Sankoff, D: Parametric genome rearrangement. Gene **172** (1996) GC11–GC17
15. Felsenstein, J.: Inferring phylogenies. Sinauer Associates (2003)
16. Hein, J., Wiuf, C., Knudsen, B., Moller, M.B., Wibling, G.: Statistical alignment: Computational properties, homology testing and goodness-of-fit. J. Mol. Biol. **203** (2000) 265–279
17. Larget, B., Simon, D.L., Kadane, B.J.: Bayesian phylogenetic inference from animal mitochondrial genome arrangements. J. Roy. Stat. Soc. B. **64(4)** 681–695
18. York, T.L., Durrett, R., Nielsen, R.: Bayesian estimation of inversions in the history of two chromosomes. J. Comp. Biol. **9** (2002) 808–818
19. Durrett, R., Nielsen, R., York, T.L.: Bayesian estimation of genomic distance. Genetics **166** (2004) 621–629
20. Miklós, I.: MCMC Genome Rearrangement. Bioinformatics **19** (2003) ii130–ii137
21. Miklós, I., Ittzés, P., Hein, J.: ParIS genome rearrangement server. Bioinformatics (2004) advance published, doi:10.1093/bioinformatics/bti060
22. Boore, J.L.: The duplication/random loss model for genome rearrangement exemplified by mitochondrial genomes of deuterostome animals. In: Sankoff, D., Nadau, J.H. (eds.): Comparative Genomics. Computational Biology Series **1** (2000) Kluwer Academic Publishers
23. Miklós, I., Lunter, G.A., Holmes, I.: A 'long indel' model for evolutionary sequence alignment. Mol. Biol. Evol. **21(3)** (2004) 529–540
24. Metropolis, N., Rosenbluth, A.W., Rosenbluth, M.N., Teller, A.H., Teller, E.: Equations of state calculations by fast computing machines. J. Chem. Phys. **21(6)** (1953) 1087–1091
25. Liu, J.S.: Monte Carlo strategies in scientific computing. Springer Series in Statistics, New-York. (2001)
26. Hastings, W.K.: Monte Carlo sampling methods using Markov chains and their applications. Biometrika **57(1)** (1970) 97–109
27. von Neumann, J.: Various techniques used in connection with random digits. National Bureau of Standards Applied Mathematics Series **12** (1951) 36–38.
28. Nadau, J.H., Taylor, B.A.: Lengths of chromosome segments conserved since divergence of man and mouse. PNAS **81** (1984) 814–818

The Distribution of Inversion Lengths in Bacteria

David Sankoff[1], Jean-François Lefebvre[2], Elisabeth Tillier[3],
Adrian Maler[1], and Nadia El-Mabrouk[2]

[1] Department of Mathematics and Statistics, University of Ottawa,
585 King Edward Avenue, Ottawa K1N 6N5, Canada
`sankoff@uottawa.ca`

[2] Département d'informatique et recherche opérationnelle, Université de Montréal,
CP 6128 succ. Centre-ville, Montréal, Québec H3C 3J7 Canada
`{lefebvre,mabrouk}@iro.umontreal.ca`

[3] Ontario Cancer Institute, Princess Margaret Hospital,
620 University Avenue, Suite 703, Toronto, Canada
`e.tillier@utoronto.ca`

Abstract. The distribution of the lengths of genomic segments inverted during the evolutionary divergence of two species cannot be inferred directly from the output of genome rearrangement algorithms, due to the rapid loss of signal from all but the shortest inversions. The number of short inversions produced by these algorithms, however, particularly those involving a single gene, is relatively reliable. To gain some insight into the shape of the inversion-length distribution we first apply a genome rearrangement algorithm to each of 32 pairs of bacterial genomes. For each pair we then simulate their divergence using a test distribution to generate the inversions and use the simulated genomes as input to the reconstruction algorithm. It is the comparison between the algorithm output for the real pair of genomes and the simulated pair which is used to assess the test distribution. We find that simulations based on the exponential distribution cannot provide a good fit, but that simulations based on a gamma distribution can account for both single-gene inversions and short inversions involving at most 20 genes, and we conclude that the shape of latter distribution corresponds well to the true distribution at least for small inversion lengths.

1 Introduction

The study of genome rearrangement has made it clear that the lengths of the chromosomal segments inverted, transposed or reciprocally translocated is not determined simply by a random choice of two breakpoints anywhere in the genome. While this is very-well documented in eukaryotes [2, 10, 13, 5], it is also true that prokaryotic genome rearrangement also operates under a variety of constraints on inversion site and length of inverted segments [12, 17, 11]. Incorporating information on such constraints into procedures for reconstructing genome divergence, e.g. in terms of weights in a parsimony analysis, probabilities in a likelihood analysis or priors in a Bayesian analysis, is a desirable goal for

J. Lagergren (Ed.): RECOMB 2004 Ws on Comparative Genomics, LNBI 3388, pp. 97–108, 2005.
© Springer-Verlag Berlin Heidelberg 2005

evolutionary methodology. With this motivation, in this paper we study the distribution of lengths of the segments that are inverted in the evolutionary history of bacterial genomes. Inherent in this study are many assumptions, not the least of which is that the distribution in question exists, i.e., represents a tendency relatively fixed over time and across the phylogenetic spectrum of bacteria. While we cannot resolve such a far-reaching question here, our results will provide a measure of confirmatory justification.

Another assumption is that inversion is the dominant process of gene order change in bacteria. Our approach will control for changes in genome size through gene gain and gene loss, but not for the effects of simply transposing segments from one area of the genome to another. This does not seem to be unwarranted; we find no systematic discussion of a transposition process in the literature on bacterial genomes, though transposition of small segments is very common in eukaryotic nuclear genomes [5, 10], and duplication-loss, which has the same effect as transposition, is often cited as an explanation for gene-order change in eukaryotic organelle genomes [3].

In a previous study [11], we analyzed the inversion lengths inferred between each of four pairs of bacterial genomes and discovered an unexpectedly high number of short inversions, single-gene inversions in particular. This contrasted with the null hypothesis that the two breakpoints of an inversion occur randomly and independently within the genome of length n, which predicts a uniform distribution $U[1, \frac{n}{2}]$ of inversion lengths, where the $\frac{n}{2}$ reflects the fact that for a circular genome, an inversion of length l is indistinguishable from the complementary inversion of length $n - l$.

The present paper builds on the previous work in two ways. First, we greatly expand our sample of genome pairs, from four to 32, deliberately picked to represent the range between closely-related and phylogenetically distant pairs, and we use a more systematic method than in the previous paper for validating relations of orthology within each pair. Second, rather than just reject the uniform null hypothesis, we attempt to pin down aspects of the probability distribution of inversion length in bacterial evolution. More precisely, we focus on the shape of this distribution only where the inversions are short, namely single-gene inversions and inversions of at most 20 genes. This rather restrained ambition is warranted by the discovery in [11], summarized in Section 2 below, that in genomes that have been even moderately rearranged by the accumulation of inversions, parsimonious methods such as that Hannenhalli-Pevzner (HP) algorithm [7], can only recover the details of very short inversions. Simulations in [11] showed that the longer inversions "recovered" by such algorithms are overwhelmingly different from those used to generate the genomes.

In the next section of this paper we recap only the part of [11] which deals with signal decay. In Sections 3 and 4 we describe our methods and data. Section 5 contains our results, which show that a two-parameter distribution function, such as the gamma distribution, is necessary to reasonably fit the numbers of short inversions observed in the 32 pairs of genomes, but that a one-parameter distribution, such as that of the negative exponential distribution, is inadequate. These results are further discussed in Section 6.

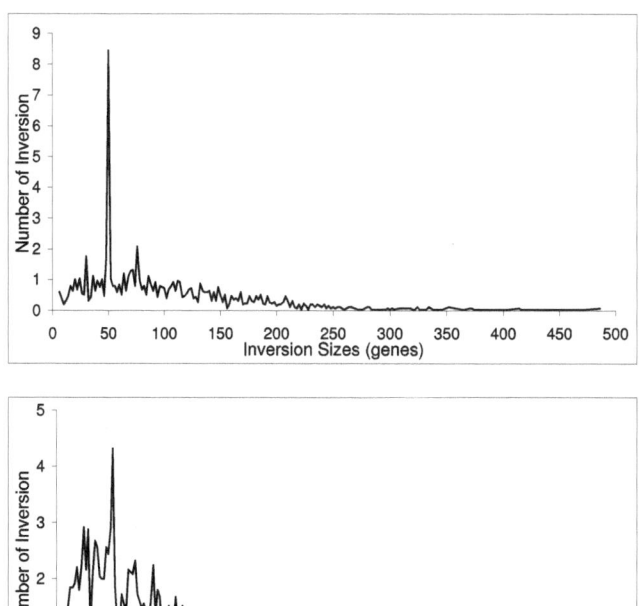

Fig. 1. Frequency of inversion sizes (or lengths) inferred by the algorithm for random genomes obtained by performing i inversions of length $l = 50$. The figure on the top is for $i = 80$ and the bottom one is for $i = 200$

2   Decay of Evolutionary Signal with Inversion Length

Consider two genomes containing the same set of genes but in different orders, where this difference is generated by evolutionary operations of a given type, such as inversions. We first ask to what extent the evolutionary histories reconstructed by the HP type of algorithm [7] actually reflect the true events. It is well-known that past a threshold of θn, where n is the number of genes and θ is in the range of $\frac{1}{3}$ to $\frac{2}{3}$, the *number* of operations begins to be underestimated by edit operation-based inferences (e.g., [8, 9]). Before that threshold, the total number may be accurately estimated but whether any signal is conserved as to the actual individual operations themselves, and which ones, is a different question.

In [11], we carried out the following test: For a genome of size $n = 1000$, we generated i inversions of length $l = 5, 10, 15, 20, 50, 100, 200$ at random, and then reconstructed the optimal inversion history, for a range of values of i. Typically, for small enough values of i, the algorithm reconstructs the true inversion history. Depending on l, however, above a certain value of i, the reconstructed inversions manifest a range of lengths, as illustrated in Figure 1 (reproduced from [11]).

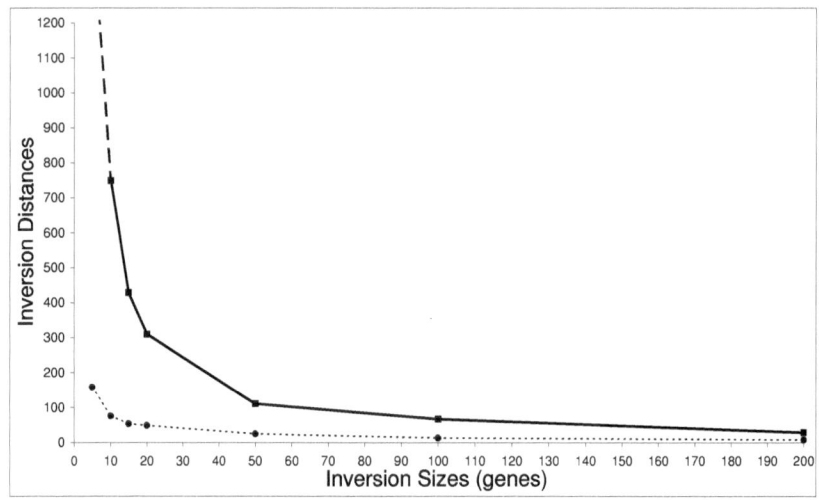

Fig. 2. The solid line plots s, and the dotted line plots r (see text above)

For each l, we calculated
$$r_l = \min\{i | \text{reconstruction has at least 5\% error}\}$$
and
$$s_l = \max\{i | \text{reconstruction has at most 95\% error}\},$$
where any inversion having length different from l is considered to be an error. Figure 2 (reproduced from [11]) plots r and s as a function of l and shows how quickly the detailed evolutionary signal decays for large inversions. Nevertheless, we note that for very small inversions, there is a clear signal preserved long after longer ones have been completely obscured.

3   Method

In our quest for the distribution of inversion lengths in bacteria, there are three steps applied to each pair of genomes in our sample:

- We use a carefully validated method for establishing orthologies between the two genomes, based on both sequence and genomic context [4].
- We calculate the inversion distance between the two genomes, as well as a number of detailed evolutionary scenarios exemplifying this distance
- We simulate a matching pair of genomes whose divergence is based on whatever distribution we are testing.

These steps are detailed in the following paragraphs.

3.1   Orthology

In the new method developed in [4], potential orthologs are evaluated according to a number of criteria:

- status of BLAST match; whether it is the best match in both directions
- quality of BLAST match; in terms of statistical significance
- scope of BLAST match with respect to the total length of the gene
- presence or absence of contextual markers conserved in both genomes
- whether there are near optimal competing genes in either genome

This enabled us to construct a matched set of orthologous genes in both genomes with a maximum of confidence. Of course, some of the matches are less clear than others, and the matches in closely related genomes tend to be less ambiguous than in distant pairs. Nevertheless these matches represent a systematic, multi-criterion, best estimate.

Once the matches are established, we constructed reduced genomes of equal length by deleting those genes not identified as being in an orthologous match. This paper reports on the analysis of these reduced genomes only, though we have also analyzed the full genomes using an inversion/insertion/deletion procedure [6]. Results from the latter were generally less clear, though they did not conflict with the results reported here.

Note that our use of reduced genomes means that our characterizations of inversions as "single-gene" or "1-20 genes" in the comparison of the reduced genomes may sometimes refer to somewhat larger inversions when the deleted genes from the unreduced genomes are restored.

3.2   Algorithm

The results of genome rearrangement algorithms are highly non-unique; many different evolutionary scenarios have the same, minimal, number of steps.

In a previous publication [1] we developed a general method that allows a choice among equally optimal solutions (i.e., the same minimal number of operations) generated by a HP type of algorithm, based on any one of many possible secondary criteria. This takes advantage of the many equally valid choices that may be available at each step of the algorithm.

Given our interest in short inversions, we adopt inversion length as our secondary criterion. Thus a solution can be obtained by selecting, at random, one of the shortest allowable inversions at each step of the HP procedure. Running the algorithm several times gives rise to several possible solutions. We can then tabulate how many times inversions of a particular length recur in the set of solutions. In [11], we showed that this length-based strategy enhanced the difference between pairs of real genomes and simulated pairs where the inversion lengths were sampled from the $U[1, \frac{n}{2}]$ distribution. The number of reconstructed single-gene and other short inversions, already higher in the real genome comparison than in the simulations, based on HP with no secondary strategy, increased markedly under the length-based strategy. There was little increase in the number of reconstructed single-gene and other short inversions for the genomes created with uniformly generated inversions. In other words, the increased number of short inversions inferred by length-based strategy was not simply an artifact of this strategy since it had little if any effect on the simulated genomes. Rather we attributed it to better detection of bonafide short inversions whose signal we know to be conserved despite extensive genome rearrangement.

3.3   Simulations

To estimate the shape of the probability distribution of inversion lengths l, we explored

– a single parameter distribution, namely a negative exponential distribution

$$p(l) = \lambda e^{-\lambda l}. \tag{1}$$

– a two-parameter distribution, namely a gamma distribution

$$p(l) = \frac{l^{\alpha-1} e^{-l/\beta}}{\beta^{\alpha} \Gamma(\alpha)}. \tag{2}$$

For each each distribution p with cumulative P, we derived simulated pairs of genomes to compare with each of the 32 real ones as follows. For a given pair of bacterial genomes, let n be the length of the reduced genome, and let i be the number of inversions necessary to derive one from the other, as measured by the HP algorithm. We sampled somewhat more than i inversions (to compensate for the bias introduced by parsimonious reconstruction in a later step) from the probability distribution and used these to evolve a new circular genome starting from $1, 2, \cdots, n$. One of the breakpoints for each inversion was located randomly on the genome, and the second was located according to the sample inversion length. If an inversion was longer than $n/2$, we discarded it and did not count it, since and inversion of length l is the same as an inversion of length $n - l$ for a circular genome. So the effective length distribution was actually

$$p^*(l) = \frac{p(l)}{P(\frac{n}{2})}$$

for

$$0 < l \leq \frac{n}{2} \tag{3}$$

and zero elsewhere.

4   The Pairs of Bacterial Genomes

We informally sampled 32 pairs of genomes from those treated in [4], choosing some that are as phylogenetically distant as possible, and some that are relatively closely related. These are listed in Table 1, which also lists i, the minimum number of inversions necessary to convert one (reduced) genome to another, the size of the reduced genome n, i.e., the number of orthologous gene pairs in the two genomes as determined by the method in [4], the normalized inversion distance i/n, and the number of single-gene and 1-20 gene inversions.

Table 1. Pairs of bacterial genomes in this study. n is the number of orthologous genes identified in the two genomes, i is the inversion distance. Pairs ordered from least to highest values of the normalized inversion distance i/n. Last two columns give the number of inversions, out of a total of i, that involve just one gene and twenty or fewer genes, respectively

genome A	genome B	i	n	i/n	1	≤ 20
Neisseria meningitidis MC58	Neisseria meningitidis Z2491	53	1606	0.03	0	3
Salmonella typhi	Shigella flexneri 2a	196	2801	0.07	4	16.4
Escherichia coli CFT073	Salmonella typhimurium LT2	244	3145	0.08	8.6	25.6
Mycobacterium leprae	Mycobacterium tuberculosis CDC1551	109	1367	0.08	4	14.8
Staphylococcus aureus Mu50	Staphylococcus epidermidis ATCC 12228	148	1805	0.08	0.2	15.8
Streptococcus agalactiae 2603	Streptococcus pyogenes	201	1156	0.17	5	28.2
Streptococcus mutans	Streptococcus pyogenes	211	1046	0.20	6	19
Agrobacterium tumefaciens C58 Uwash Circ	Sinorhizobium meliloti	347	1705	0.20	14.6	50.4
Escherichia coli CFT073	Yersinia pestis CO92	642	2363	0.27	13	71
Pseudomonas aeruginosa	Pseudomonas putida KT2440	996	3189	0.31	34	117.2
Corynebacterium glutamicum	Mycobacterium tuberculosis CDC1551	380	1087	0.35	11.4	39.4
Bacillus halodurans	Bacillus subtilis	748	1912	0.39	31	91.6
Salmonella typhi	Vibrio cholerae ChI	584	1479	0.39	20	109
Listeria innocua	Staphylococcus aureus Mu50	471	1085	0.43	22.4	69.6
Escherichia coli K12	Vibrio cholerae ChI	717	1648	0.44	29	108
Bacillus halodurans	Listeria innocua	518	1186	0.44	22.6	61.2
Bacillus halodurans	Oceanobacillus iheyensis	818	1856	0.44	18	105
Listeria monocytogenes	Staphylococcus epidermidis	486	1080	0.45	15	78
Clostridium acetobutylicum	Clostridium perfringens	564	1211	0.47	22.6	70.2
Salmonella typhimurium LT2	Shewanella oneidensis	741	1474	0.50	33.6	100
Clostridium perfringens	Thermoanaerobacter tengcongensis	427	841	0.51	23	60
Oceanobacillus iheyensis	Thermoanaerobacter tengcongensis	452	853	0.53	18.4	61.8
Pseudomonas putida KT2440	Vibrio cholerae ChI	639	1160	0.55	28.4	90.4
Staphylococcus epidermidis ATCC 12228	Thermoanaerobacter tengcongensis	359	623	0.58	17	56.6
Listeria innocua	Thermoanaerobacter tengcongensis	450	770	0.58	21.8	64.2
Clostridium perfringens	Staphylococcus epidermidis ATCC 12228	391	640	0.61	19.4	60
Bacillus halodurans	Clostridium perfringens	619	944	0.66	28	81
Mycobacterium tuberculosis CDC1551	Mycoplasma penetrans	90	137	0.66	6	16
Bacillus subtilis	Streptococcus agalactiae 2603	533	806	0.66	27.8	65
Bacillus halodurans	Streptococcus pneumoniae R6	509	737	0.69	21.2	90.4
Staphylococcus aureus Mu50	Streptococcus pyogenes MGAS8232	649	939	0.69	25	127
Streptococcus mutans	Thermoanaerobacter tengcongensis	432	598	0.72	18	76

We note that parts of the evolutionary history separating many of the gene pairs are shared; perhaps the most obvious example is the *E.coli – Vibrio* and *Salmonella – Vibrio* comparisons, since these reflect a largely similar historical divergence, *E.coli* and *Salmonella* having a relatively recent common ancestor. This kind of dependence, which in general increase measures of dispersion but not bias, is not as great among our other pairs of genomes, and is in any case virtually impossible to avoid in a phylogenetic context.

5   Results

Applying our algorithm to the 32 pairs of bacterial genomes, repeating each comparison ten times with different random choices of shortest allowable inversion at each step, we counted the average number of single-gene inversions and the average number of inversions of length 20 or less. These were normalized by n and plotted against the normalized inversion distance i/n in Figure 3. We also plotted on Figure 3 the result of our simulations based on the negative exponential and gamma distributions. For the negative exponential, it can be seen that a value of λ that allows the curve obtained from $p(l) \leq 1$ to fit the real data on single-gene inversions does not allow the curve obtained from $p(l) \leq 20$ to fit the real data on inversions of length 20 and less, and vice-versa. For the gamma distribution, on the other hand, values of α and β can be found that fit both sets of data, although for 1-20 gene inversions, the fit breaks down when $i > n/2$.

We found such values of the parameters of the gamma distribution by minimizing the sum of squared differences, between each real pair of genomes and the corresponding simulated pair, of the normalized number of single-gene inversions in a minimal inversion scenario in plus the analogous difference for the normalized number of 1-20 gene inversions. The latter differences were weighted by a factor of 0.1, since the number of short inversions was approximately 10 times as large as the number of single-gene inversions. We iterated by fixing each parameter in turn and searching for the minimizing value of the other parameter.

6   Discussion

To what extent do our results bear on the question of whether there is a universal distribution of inversion lengths across the bacterial domain? After all, this distribution is the result of numerous mechanistic mutational processes at the chromosomal level as well as selective processes operating on cell form and function, both of which can be expected to vary among genomes.

The generality of the distribution can be assessed in part by the deviation of the sample points from the overall trend in Figure 3. While it is true there is a degree of statistical fluctuations, our results are thoroughly compatible with the hypothesis that all the pairs are following a common tendency. That the more distantly related genome pairs have fewer 1-20 gene inversions than the corresponding simulated pairs indicates some tendency for the signal from the short inversions to be lost for reasons other than genome rearrangement, which should affect the simulated and real pairs in the same way. The observed shortfall in the number of short inversions for normalized distances greater than about 0.45 is partly due to an greater incidence of undetectable orthology in the more distant pairs, and partly to our way of treating unequal gene complements, of accumulated gene gain and loss for these pairs. Neither of these problems affect the simulated genome pairs. Whichever the explanation, the fact remains that all the distant pairs manifest the same shortfall, and there is no idiosyncratic behaviour from genome pair to genome pair evident at the aggregate level. Note that overall inversion *frequency* is not addressed in our analysis, since we are using no external time measure to calibrate the genomic distances, but this is not pertinent to our results.

Recently, attention has been drawn to the prevalence and significance of short inversions, albeit more in eukaryotes [2, 10, 16, 13, 5] than in prokaryotes [15, 11]. Here we have advanced our approach to the study of short inversions, taking advantage of the greatly elevated persistence in their evolutionary signal, compared

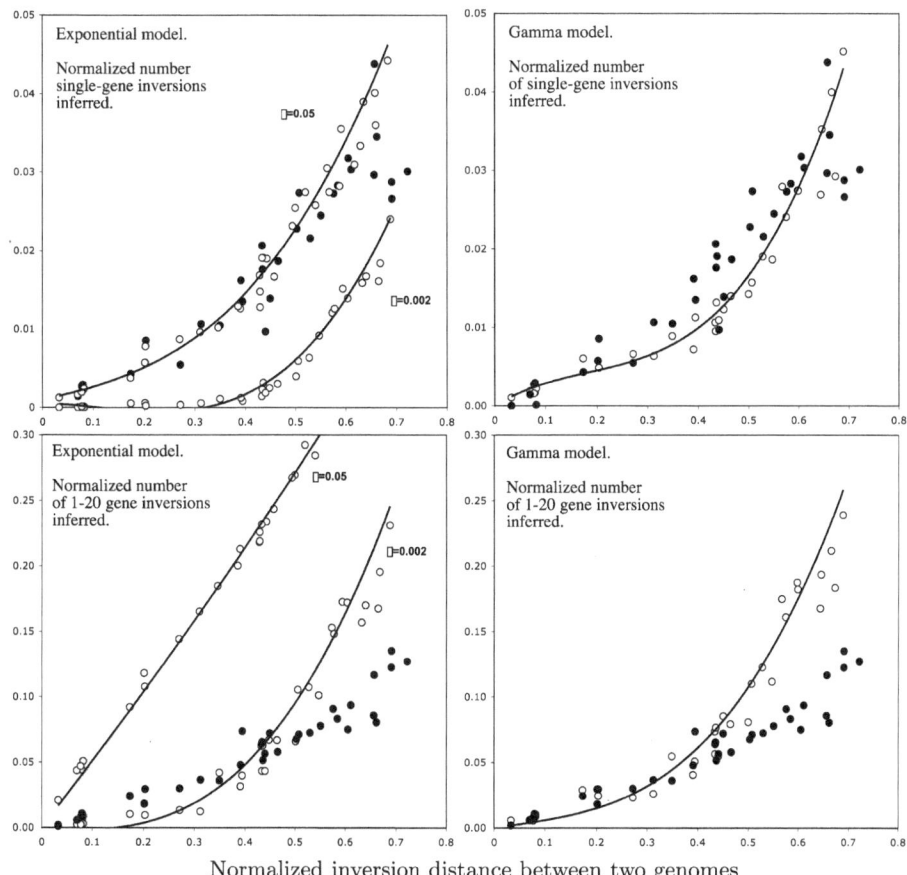

Fig. 3. Fit of exponential and gamma models (open dots and trend lines) to data on single gene inversions and 1-20 gene inversions (filled dots). Exponential parameter $\lambda = 0.002$ or 0.05, gamma parameters $\alpha = 0.60, \beta = 1200$

to that of longer inversions. We found that the distribution of *inferred* inversion lengths could be accounted for by a gamma distribution for the *generating* inversions, with a high proportion of single-gene and other short inversions and a rapid but non-exponential initial decline. The initial 30 values of the gamma distribution with parameters $\alpha = 0.60, \beta = 1200$ are depicted in Figure 4. Note that we do not consider any but this first few values of l. The upper tail of the gamma distribution is not relevant to this study; indeed our generation procedure truncates most all of the domain of the distribution greater than $\frac{n}{2}$. In any case, we are using the gamma as a descriptive device and are not suggesting it is theoretically privileged in being mathematically derived from some mutational or selective model for the inversion process. Note that in [11] we ruled out a uniform distribution as descriptively inadequate, and in the present paper we also ruled out the exponential distribution.

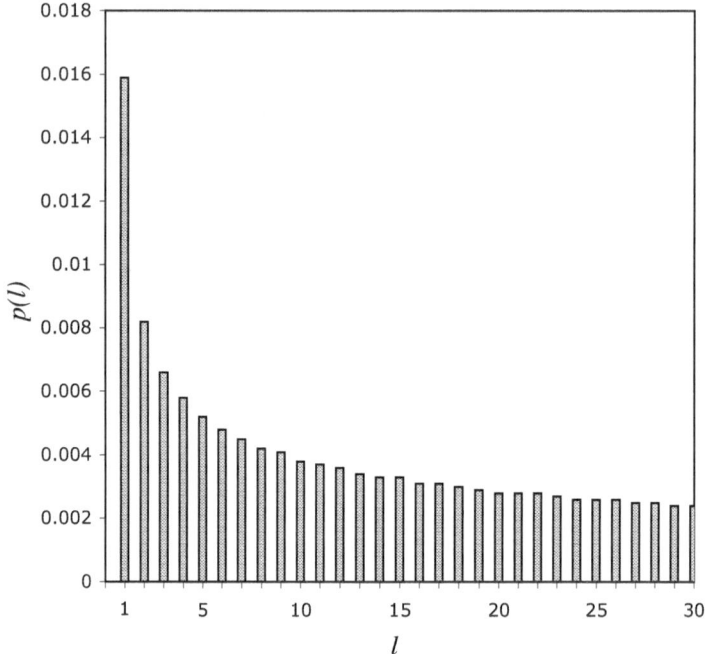

Fig. 4. Gamma distribution with parameters $\alpha = 0.60, \beta = 1200$

How can the preference for short inversions be explained? We suggest that it is a combination of factors:

- Single-gene inversions may represent a particular evolutionary mechanism with selective functional consequences. They may allow a gene to obtain transcriptional independence from its erstwhile operon, or to otherwise change its expression pattern, or to take advantage of new or altered functionality, or to participate in a different pathway through a more appropriate genomic positioning (cf genomic hitchhiking [14]).
- Single-gene inversions may simply be the clearest manifestation of a universal tendency towards short inversions as the least disruptive of the gene proximity configuration, and attendant functionality, of a genome. In [15], we argued that a predisposition for such inversions in small genomes might explain the prevalence of internally-shuffled "gene clusters" found across many sequenced genomes in microorganisms, in contrast to the "conserved segments", including fixed gene order, pattern characteristic of the higher eukaryotes.
- Mechanistic process that favour mutational processes operating over short distances.

Any knowledge about the distribution of inversion lengths would be invaluable to the inference of genome rearrangements. It is very difficult to obtain suitable data, however, so that the approach offered here is an example of the indirect methods that must be developed in order to eventually home in on the true distribution.

Acknowledgments

Research supported by grants from the Natural Sciences and Engineering Research Council (NSERC), and the *Fonds québécois de recherche sur la nature et les technologies*. DS holds the Canada Research Chair in Mathematical Genomics. NE-M, ET and DS are affiliated with the Evolutionary Biology Program of the Canadian Institute for Advanced Research. We thank the anonymous referees for extensive comments and suggestions on an earlier version of this paper.

References

1. Ajana, Y., Lefebvre, J.F. , Tillier, E. and El-Mabrouk, N. (2002). Exploring the set of all minimal sequences of reversals - An application to test the replication-directed reversal hypothesis. *Algorithms in Bioinformatics, Second International Workshop, WABI*, Guigó,R. and Gusfield,D. Eds., Lecture Notes in Computer Science, **2452**, 300–15. Springer Verlag.
2. Bennetzen, J.L. and Ramakrishna, W. (2002) Numerous small rearrangements of gene content, order and orientation differentiate grass genomes. *Plant Molecular Biology*, **48**, 821–7.
3. Boore, J.L (2000) The duplication/random loss model for gene rearrangement exemplified by mitochondrial genomes of deuterostome animals, *Comparative Genomics*, Sankoff,D. and Nadeau,J.H. Eds., 133–47. Dordrecht, NL: Kluwer Academic Press
4. Burgetz, I.J., Shariff, S., Pang, A. and Tillier, E.R.M. (2004). Positional homology in bacterial genomes. Submitted ms.
5. Coghlan, A. and Wolfe, K.H. (2002) Fourfold faster rate of genome rearrangement in nematodes than in Drosophila. *Genome Research*, **12**, 857-67.
6. El-Mabrouk, N. (2001) Sorting signed permutations by reversals and insertions/deletions of contiguous segments. *Journal of Discrete Algorithms*, **1**, 105–122.
7. Hannenhalli, S. and Pevzner, P. A. (1999). Transforming cabbage into turnip (polynomial algorithm for sorting signed permutations by reversals). *Journal of the ACM*, **48**, 1–27.
8. Kececioglu, J. and Sankoff, D. (1994) Efficient bounds for oriented chromosome-inversion distance. *Combinatorial Pattern Matching. Fifth Annual Symposium*, Crochemore, M. and Gusfield, D., Eds., Lecture Notes in Computer Science **807**, 307–25. Springer Verlag.
9. Kececioglu, J. and Sankoff, D. (1995) Exact and approximation algorithms for sorting by reversals, with application to genome rearrangement. *Algorithmica*, **13**, 180-210.
10. Kent, W. J., Baertsch, R., Hinrichs, A., Miller, W. and Haussler, D. (2003). Evolution's cauldron: Duplication, deletion, and rearrangement in the mouse and human genomes. *Proceedings of the National Academy of Sciences, USA*, **100**, 11484–9.
11. Lefebvre, J.-F., El-Mabrouk, N., Tillier, E. and Sankoff, D. (2003). Detection and validation of single-gene inversions. *Bioinformatics*, **19**, i190–6.

12. Mahan, M.J. and Roth, J.R. (1991) Ability of a bacterial chromosome segment to invert is dictated by included material rather than flanking sequence. *Genetics*, **129,** 1021-32.

13. McLysaght, A., Seoighe, C. and Wolfe, K.H. (2000). High frequency of inversions during eukaryote gene order evolution. *Comparative Genomics*, Sankoff,D. and Nadeau,J.H. Eds.,47–58. Dordrecht, NL: Kluwer Academic Press

14. Rogozin, I.B., Makarova, K.S., Murvai, J., Czabarka, E., Wolf, Y.I., Tatusov, R.L., Szekely, L.A. and Koonin, E.V. (2002) Connected gene neighborhoods in prokaryotic genomes. *Nucleic Acids Research*, **30,** 2212–23.

15. Sankoff, D. (2002). Short inversions and conserved gene clusters. *Bioinformatics*, **18,** 1305–1308.

16. Thomas, J. W, and Green, E. D. (2003). Comparative sequence analysis of a single-gene conserved segment in mouse and human. *Mammalian Genome*, **14,** 673–8.

17. Tillier, E.R.M. and Collins, R. (2000) Genome rearrangement by replication-directed translocation. *Nature Genetics*, **26,** 195–7.

Estimators of Translocations and Inversions in Comparative Maps

David Sankoff[1] and Matthew Mazowita[1]

Department of Mathematics and Statistics, University of Ottawa,
585 King Edward Avenue, Ottawa ON, Canada, K1N 6N5
{sankoff,mmazo039}@uottawa.ca
http://albuquerque.bioinformatics.uottawa.ca

Abstract. In a comparative map, the number of translocations in the evolutionary history of a chromosome can be estimated solely on the basis of the conserved *syntenies* it contains. This estimate, subtracted from the number of conserved *segments*, then allows the estimation of the number of inversions that have affected the chromosome. Summing these estimates over all chromosomes provides a startlingly accurate estimator (as assessed by simulation) of the total number of rearrangements of each type occurring in the evolutionary divergence of two genomes.

1   Introduction

The quantitative comparative study of whole-genome maps, exemplified by the linkage-based work of Nadeau and Taylor [8] and more recent versions based on gene content of conserved segments [16, 15, 5], makes no formal reference to the processes that create the breakpoints between conserved segments while progressively fragmenting these segments. It only assumes implicitly that the number of breakpoints and segments increases in proportion to the number of rearrangement events affecting either of the two genomes being compared. In contrast, the algorithmic approach to genome rearrangements [4, 19] infers a most parsimonious history of specific inversions and reciprocal translocations. The particulars of this inference are not always reliable due to the highly non-unique nature of the solutions, the characteristic underestimation of parsimony and, especially, the susceptibility of these methods to lose the details, though not the overall trends, in the evolutionary signal [17].

Between these analytical extremes of ignoring the processes giving rise to a comparative map and ambitiously trying to infer them in all their detail, are there any limited aspects of rearrangement history that can be inferred with some confidence? Building on the ideas in [13] and [14], we claim that by analyzing the number of conserved syntenies in a comparative map, we can accurately infer, using a simple estimator, the number of reciprocal translocations involved in generating this map. Furthermore, depending on the degree of resolution of the map, by contrasting the number of conserved segments with the number of conserved syntenies, we can estimate the number inversions or other intra-chromosomal events. Applying this methodology to data compilations on human-mouse maps at differing levels of resolution shows that the estimated number

J. Lagergren (Ed.): RECOMB 2004 Ws on Comparative Genomics, LNBI 3388, pp. 109–122, 2005.
© Springer-Verlag Berlin Heidelberg 2005

of reciprocal translocations is relatively stable, but that the inferred number of inversions increases with finer resolution.

2   Models

We can model the autosomes of a genome as c linear segments with lengths $p(1), \cdots, p(c)$, proportional to the number of base pairs they contain, where $\sum_{i=1}^{c} p(i) = 1$. To balance realism and simplicity in our model, we:

- Set aside the sex chromosomes, which are largely excluded from inter-chromosomal exchanges.
- Impose a threshold and a cap on chromosome size, rejecting any translocation that results in a chromosome too small or too large. Theories about meosis, e.g. [18], can be adduced for these constraints, though there are clear exceptions, such as the "dot" chromosomes of avian and some reptilian and other vertebrate genomes [2, 1].
- Impose a left-right orientation on each chromosome, such that a left-hand fragment must always rejoin a right-hand fragment. In reality, an inverted left-hand fragment may rejoin another left-hand fragment, but the only statistical effect of our restriction is to ensure that, throughout the simulation, each chromosome always retains a segment, however small it may become, containing its original left-hand extremity. This restriction models the conservation of the centromere without introducing complications such as trends towards or away from acrocentricity.
- Postpone our consideration of chromosome fusion and fission, so that the number of chromosomes is constant throughout the time period governed by the model. Later, we simply assume that the case where fusions or fissions occur will be well approximated by interpolating two models (with fixed chromosome number) corresponding to the two genomes being compared.

We also assume the two breakpoints of a translocation are chosen independently according to a uniform distribution over all autosomes, conditioned on their not being on the same chromosome. There is no statistical evidence [11] that translocational breakpoints cluster in a non-random way on chromosomes, except in a small region immediately proximal (within 50-300Kb) to the telomere in a wide spectrum of eukaryote lineages [7].

A reciprocal translocation between two chromosomes h and k consists of breaking each one, at some interior point, into two segments, and rejoining the four resulting segments such that two new chromosomes are produced, each containing a left-hand part of one of the original chromosomes and the right-hand part of the other. We label each new chromosome according to which left-hand it contains, but for each of its constituent segments, we retain the information of which ancestral chromosome it derived from.

With further translocations, if a breakpoint falls into a previously created segment on chromosome i, it divides that segment into two new segments, the left-hand one remaining in chromosome i, while the right-hand one, and all the other segments to the right of the breakpoint, are transferred to the other chromosome involved in the translocation,

In contrast to translocations, the two breakpoints of an inversion cannot be considered to be independently positioned on the chromosome. According to Kent et al. [6] the median length of an inversion, of which there are many thousands in the mouse-human comparison, is less than 1 Kb, so that on a chromosomal scale most inversions have their two breakpoints very close together.

There are no definitive data on the distribution of inversion lengths. The best that exist, based on human-mouse comparisons, were published in Table 2 of [6], which estimates about 8000 inversions (with or without partial or complete duplications) with median length from 300-800 bp (depending on the subcategory of inversion), including 160 inversions longer than 100 Kb within longer syntenic blocks (*intrablock* inversions). In [10] it is estimated that there are 150 inversions longer than 1 Mb; these each involve at least one whole syntenic block (*suprablock* inversions) and hence do not overlap with the 160 previously mentioned. These data (median = 600 with 310 inversions out of 8000 longer than 1 Mb) determine a gamma distribution with shape parameter $\alpha = 6.539$ and scale parameter (on a logarithmic scale) $\beta = 0.447$. This distribution has a median of 600 bp and a 0.05 tail > 100 Kb, to include 0.02 intrablock, 0.02 suprablock and a generous 0.01 for those inversions falling through the cracks between the definitions of 100 Kb intrablock and 1Mb suprablock inversions.

In our simulations, we will test our models with a much more disruptive distribution of inversion lengths. We generate these lengths according to a gamma distribution with shape parameter $\alpha = 3$ and scale parameter (on a logarithmic scale) $\beta = 1.127$.

One inversion breakpoint is chosen at random on the genome, as with translocations, and the second is chosen equiprobably to the left or right and according to the specified gamma. If the second breakpoint exceeds the end of the chromosome, both breakpoints are disregarded and a substitute inversion is generated. With this truncation protocol, keeping in mind the logarithmic scale, only about 7 % of the inversions are larger than 1 Mb. For any accounting of rearrangements which allows 8000 or more inversions of all sizes, seven percent still represents a generous number of very large inversions (> 1 Mb). We do this to provide as severe as possible a test of the formulae we will develop, which are more susceptible to fail if there are many large inversions.

The total number of segments on a human chromosome i is

$$n^{(i)} = t^{(i)} + 2u^{(i)} + 1, \tag{1}$$

where $t^{(i)}$ is the number of translocational breakpoints on the chromosome, and $2u^{(i)}$ is the number of inversion breakpoints.

3   Prediction and Estimation

We assume that our random translocation process is temporally reversible, and to this effect we show in Figure 1 and Section 4.1 that the equilibrium state of our process well approximates the observed distribution of chromosome lengths in the human genome. This assumption allows us to treat the mouse genome as ancestral and the human as derived (or vice versa), instead of considering them as diverging independently from a common ancestor.

To start with, we consider a process that involves translocations but no inversions. At the outset, assume the first translocation on the human lineage involves ancestral chromosome i. The assumption of a uniform density of breakpoints across the genome implies that the "partner" of i in the translocation will be chromosome j with probability $p_i(j) = \frac{p(j)}{1-p(i)}$. Thus the probability that the new chromosome labelled i contains no fragment of ancestral chromosome j, where $j \neq i$, is $1 - p_i(j)$. For small $t^{(i)}$, after chromosome i has undergone $t^{(i)}$ translocations, the probability that it contains no fragment of the ancestral chromosome j is approximately $(1 - p_i(j))^{t^{(i)}}$, neglecting second-order events, for example, the event that j previously translocated with one or more of the $t^{(i)}$ chromosomes that then translocated with i, and that a secondary transfer to i of material originally from j thereby occurred.

Then the probability that the new (i.e., human) chromosome i now contains at least one fragment from j is approximately $1 - (1 - p_i(j))^{t^{(i)}}$ and the expected number of ancestral chromosomes with at least one fragment showing up on human chromosome i is

$$E(c^{(i)}) \approx 1 + \sum_{j \neq i} [1 - (1 - p_i(j))^{t^{(i)}}], \qquad (2)$$

where the leading 1 counts the fragment containing the left-hand endpoint of the ancestral chromosome i itself. We term $c^{(i)}$ the number of *conserved syntenies* on chromosome i.

3.1   The Case of No Saturated Chromosomes

Substituting $c^{(i)}$ for $E(c^{(i)})$ in eqn (2) suggests solving

$$c - c^{(i)} = \sum_{j \neq i} [1 - p_i(j)]^{\widehat{t^{(i)}}}, \qquad (3)$$

to provide an estimator of $t^{(i)}$. Newton's method, initialized by the estimator in Section 3.4, converges rapidly for the range of parameters used in our studies, as long as $c^{(i)} \neq c$. This is the empirically interesting case; we know of no comparative map where a chromosome of one genome shares at least one significant syntenic segment with every autosome of the other genome.

Then $\hat{t} = \frac{1}{2} \sum_{i=1}^{c} \widehat{t^{(i)}}$ is an estimator of the total number of translocations intervening between the ancestral and modern genome, since each translocation is counted on two chromosomes.

In the hypothetical case $c^{(i)} = c$, we say chromosome i is saturated and there is no solution to eqn 3. For the sake of completeness, we will also study this case.

3.2   When There Are Some Saturated Chromosomes

If the genome after a certain number translocations contains at least one saturated chromosome, i.e., with $c^{(i)} = c$, our estimator must take on a different form.

Let c^* be the number of saturated chromosomes. Suppose there have been t translocations in the evolutionary history, with

$$t^{(i)} = 2tp(i) \tag{4}$$

affecting autosome i. Since the probability is approximately $1 - [1 - p_i(j)]^{t^{(i)}}$ that at least one segment from original chromosome j is contained by chromosome i, if these events were independent for all the $c - 1$ original chromosomes j, (which they are obviously are not, for small values of t), then the probability that $c^{(i)} = c$, i.e., chromosome i contains segments from all $c - 1$ of the others, as well as the original i by default, would be the

$$P(i,t) = \prod_{j \neq i} (1 - [1 - p_i(j)]^{t^{(i)}}). \tag{5}$$

Now, we may assume independence is asymptotically approached with large t, so that the expected number of saturated chromosomes $E(c^*)$ is approximately P_t, where

$$P_t = \sum_{i=1}^{c} P(i,t), \quad \sigma^2 = \sum_{i=1}^{c} P(i,t)(1 - P(i,t)). \tag{6}$$

These quantites can all be calculated from the $p_i(j)$ based on the given $p(i)$. Then, in the presence of c^* saturated chromosomes, we define $\hat{t}(c^*)$ to be the inverse function of P_t applied to c^*.

3.3   The Completely Saturated Case

The estimators in Section 3.2 are defined as long as $c^* < c$. When all c chromosomes are saturated, the *completely saturated* case, the best we can do is to define $\hat{t}(c) = \hat{t}(c - 1)$, with the understanding that this may well be a severe underestimate.

3.4   Equal Size Chromosomes

When all chromosomes are of equal size, a situation that can be maintained only by requiring the chromosomal fragments exchanged during translocation to be of the same length,

$$E(c^{(i)}) \approx 1 + (c - 1)[1 - (1 - \frac{1}{c - 1})^{t^{(i)}}]. \tag{7}$$

In this case [14], eqn (3) is directly solved as:

$$\widehat{t^{(i)}} = \frac{\log(c - 1) - \log(c - c^{(i)})}{\log(c - 1) - \log(c - 2)}. \tag{8}$$

3.5   The Effect of Inversions

Inversions have the effect of fragmenting and changing the order of the segments that are transferred among chromosomes by translocation. An estimate of the number of inversion breakpoints on a chromosome is derived from eqn (1) as

$$\widehat{2u^{(i)}} = n^{(i)} - \widehat{t^{(i)}} - 1, \tag{9}$$

and the number of inversions will be half that. The presence of inversions will have an effect on the estimate of t. Modeling this effect of inversions is not easy; prior inversions on chromosome j can either increase or decrease the effect of a translocation of i and j on $c^{(i)}$. As the inversion rate increases, however, the process becomes a "gossip" process among the c autosomes – after each interaction (i.e., communication) between two chromosomes, they both contain material from every original chromosome in the union of the two chromosomes before the interaction. Here, the chromosomes are saturated at a very rapid rate.

4   Simulations

4.1   Equilibrium Distribution of Chromosome Size

Models of accumulated reciprocal translocations for explaining the observed range of chromosome sizes in a genome date from the 1996 study of Sankoff and Ferretti [12]. They proposed a lower threshold on chromosome size in order to reproduce the appropriate size range in plant and animal genomes containing from two to 22 autosomes. A cap on largest chromosome size has also been proposed [18] and shown to be effective [3]. Economy and elegance in explaining chromosome size being less important in the present context than simulating a realistic equilibrium distribution of these sizes, we imposed both a threshold of 50 Mb and a cap of 250 Mb on the process described in Section 2, simply rejecting any translocations that produced chromosomes out of the range. These values were inspired by the relative stability across primates and rodents evident in the data in Table 1.

Simulating the translocation process 100 times up to 10,000 translocations each produced the equilibrium distribution of chromosome sizes in Figure 1. The superimposed distribution of human autosome sizes is very close to the equilibrium distribution.

Table 1. Shortest and longest chromosome, in Mb

genome	shortest	longest
mouse	61	199
human	47	246
rat	47	268
chimp	47	230

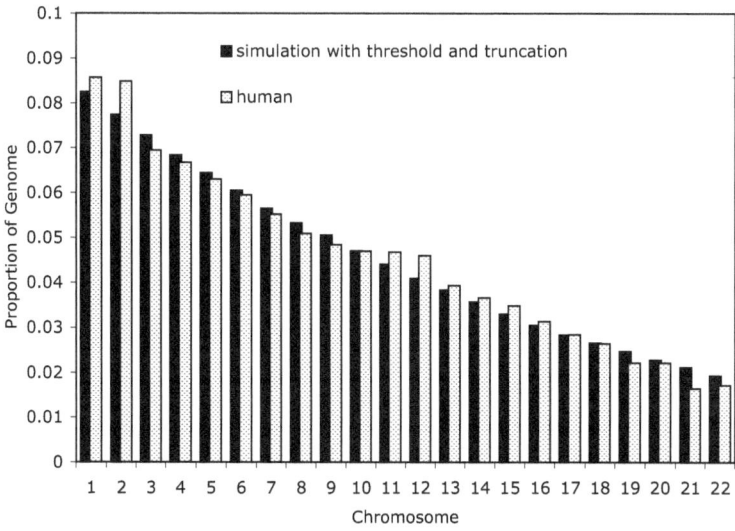

Fig. 1. Comparison of equilibrium distribution of simulated chromosome sizes with human autosome sizes

4.2   Domains of the Estimators

In 1000 runs of the simulation, the smallest t for which any chromosome was found to be saturated was 127. Indeed, up to $t = 348$, no saturated chromosomes were found in over half of the runs. By $t = 552$, however, all thousand runs had at least one saturated chromosome. Figure 2 depicts how the mean number of saturated chromosomes increases as a function of t.

As mentioned in Section 2, the range of empirical interest, at least for human-mouse comparisons, is thus well within the range of the estimator based on eqn (3). It is conceivable, however, that some remotely related genomes might require the estimator based on eqn (5).

In 1000 runs, the smallest t for which a genome was found to be completely saturated was 1747. By $t = 2749$, half of the runs were completely saturated; by $t = 3886$ all were completely saturated.

4.3   Performance of the Estimator
When There Are No Saturated Chromsomes

Figure 3 (left) depicts the estimated number of translocations as a function of the true number t in the simulation, when no saturated chromosomes are encountered. For $t < 400$, the estimator $\hat{t}$ appears completely unbiased, with only moderate variance. As explained in Section 4.2 and as can be seen in Figure 2, $t < 400$ is also the range where saturated chromosomes are rarely encountered.

For higher values of t we had to run the simulation many additional thousands of times to accumulate 100 runs without any saturated chromosome, fully

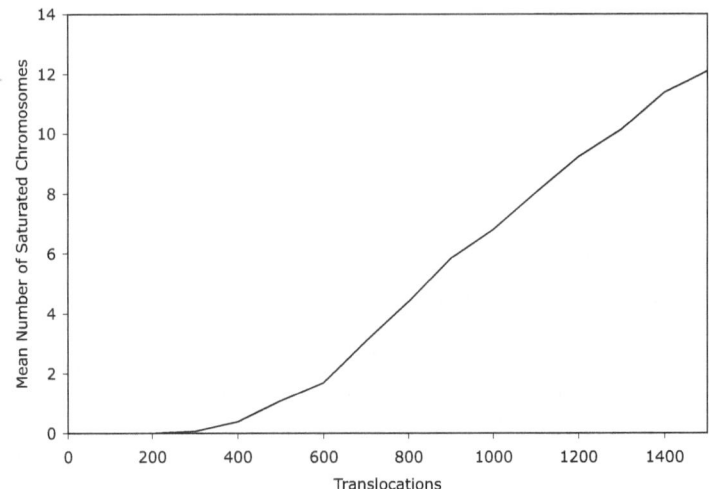

Fig. 2. Average number, over 100 runs, of saturated chromosomes per genome, as a function of the number of translocations. There are very few genomes with saturated chromosomes for $t < 400$

conscious that this atypical sub-sample was unlikely to result in accurate estimates of t. This accounts for the severe bias for large t.

4.4   Performance of the Estimator with Some Saturated Chromsomes

Application of the version of the estimator presented in Section 3.2 to instances where one or more of the chromosomes are saturated gave the results in Figure 3 (right). This has a small (around 70 translocations) constant (hence easily corrected) negative bias over the range from $t = 600$, where there is still a minority of genomes with any saturated chromosomes, to $t = 2000$, when most genomes have many saturated chromosomes. The error rate, however, is much higher than the estimator based on eqn (3) over the range where they can be compared.

For low and high values of t we had to run the simulation many extra times to accumulate 1000 runs with at least one but less than 22 saturated chromosomes. The bias this introduces is evident for $t < 500$.

4.5   The Effect of Inversions on $\hat{t}$

In our simulations, we interspersed v inversions between successive translocations, for $v = 0, 1, 10, 50$ and 100. The effect of this was to bias $\hat{t}$ positively, a proportion of inversions being inferred as translocations. This is depicted in Figure 4. This bias seems quite severe for the larger values of v, but it should be

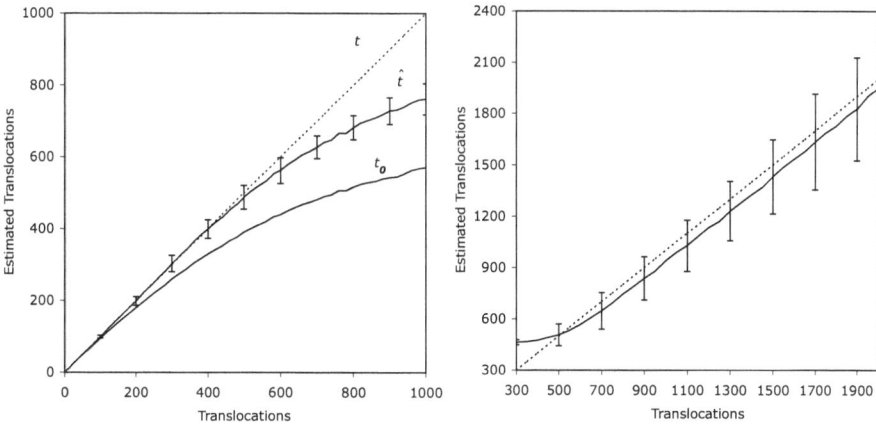

Fig. 3. (Left) Mean value, over 100 runs, of $\hat{t}$ as a function of t where there are no saturated chromosomes. Error bars ± 1 s.d. Estimator t_0 based on equal-length chromosome model, eqn (8). (Right) Mean value, over 1000 runs, of $\hat{t}$ as a function of t for the case of saturated chromosomes, using the version of the estimator in Section 3.2. Note that as in Figure 2, there are very few genomes with saturated chromosomes for $t < 400$

remembered from Section 2 that to detect these effects we are using very exaggerated inversion lengths. When we substitute a more realistic gamma function, no bias is apparent, even for $v = 50$ as indicated in the figure.

It is ironic that as while high inversion rate increases the bias in translocations rates, it actually decreases the bias in estimating the inversion rate, as a proportion of the total number of inversions. This follows since a small proportion of the number of inversions will be a large proportion of the number of translocations.

4.6   The Effect of Inversions on Saturation

The addition of inversions to the simulation accelerates the saturation of the chromosomes.

In our simulations, the lowest t's for which we encountered a saturated chromosome ($c^* = 1$) was $t = 127$ with no inversions (the second lowest was $t = 195$), $t = 140$ with one inversion per translocation, $t = 145$ with ten inversions per translocation, $t = 133$ for 50 inversions, $t = 110$ for 100 inversions, and $t = 38$ for the gossip process (1000 runs with zero, one, ten inversions and gossip, 100 runs with 50 and 100 inversions).

The lowest t's for which we encountered a completely saturated case ($c^* = c$) was $t = 1747$ with no inversions, $t = 1653$ with one inversion per translocation, $t = 1177$ with ten inversions per translocation, $t = 729$ for 50 inversions, $t = 450$ for 100 inversions, and $t = 71$ for the gossip process.

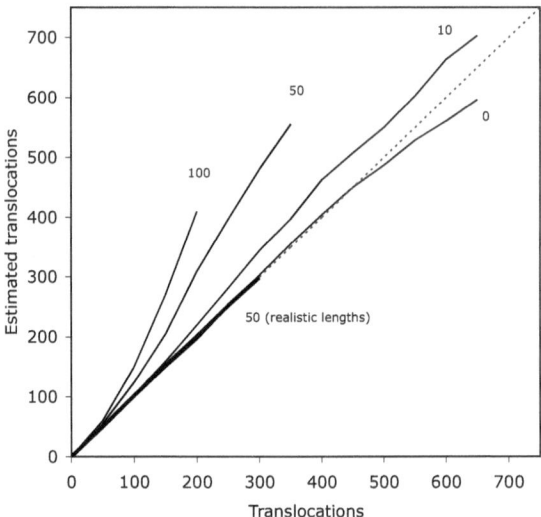

Fig. 4. Increase in number of translocations inferred as a function of intrachromosomal activity (inversions). Curves labelled by number of inversions per translocation. Curves end when computing time considerations make it unfeasible to accumulate 100 runs with no saturated chromosomes. Data are shown for simulations with two alternative models of inversions, one with a realistic inversion length distribution and the other with exaggerated inversion lengths. The heavy line shows the result of using the realistic distribution: virtually no effect on the accuracy of the estimator. The thin line shows what would happen were the inversion lengths unrealistically large

The way in which c' grows in the presence of large numbers of inversions should approach the performance of the gossip process discussed in Section 3.5, so that saturation should be reached at the same time as the gossip process. It is clear from these results that even with 100 inversions per translocation, we are still quite far from the limiting case.

5   Application to the Human-Mouse Comparison

To illustrate the use of the estimators, for the 22 human autosomes, a 100 Kb resolution construction abstracted from the UCSC Genome Browser indicates 237 autosomal segments, while the sum of the $c^{(i)}$ is 107. Solving eqn (3) for each autosome, based on its $c^{(i)}$, summing the 22 values of $\widehat{t^{(i)}}$, and dividing by 2, gives a total of 50 translocations.

By eqn (1), this leaves unaccounted for

$$2 \sum \widehat{u^{(i)}} = \sum n^{(i)} - 2\hat{t} - 22$$
$$= 115$$

segments, which must be attributed to local rearrangements such as $\hat{u} \approx 58$ inversions.

Table 2. Inference of inter- and intra-chromosomal rearrangements based on number of conserved segments and number of segment-sharing autosome pairs in the two genomes. Calculated separately from data on segments on 22 human autosomes (H) and on 19 mouse autosomes (M). Sources: UCSC Genome Browser May 2004, builds Mm5 and Hg17, level = 1, alignments $\geq$ 100 Kb, GRIMM 300 Kb and 1 Mb blocks, from http://nbcr.sdsc.edu/GRIMM/HMR_Aug2003, NCBI human build 34.3 and mouse build 32.1

resolution of comparative map	autosomal segments $\sum n^{(i)}$ H/M	segment-sharing chromosome pairs $\sum c^{(i)}$	inter-chromosomal $\frac{1}{2}\sum \widehat{t^{(i)}}$ H/M/**mean**	intra-chromosomal $\sum \widehat{u^{(i)}}$ H/M/**mean**
100 Kb (UCSC)	237 254	107	50 51 **50**	58 67 **62**
300 Kb (GRIMM)	377 377	109	50 52 **51**	127 127 **127**
1 Mb (GRIMM)	268 268	104	47 49 **48**	76 76 **76**
n/a (NCBI)	379 381	110	51 53 **52**	128 128 **128**

Table 2 shows the results of these calculations for this and a number of other maps of various levels of resolution. Of interest is the relative stability of the estimates of the number of reciprocal translocations versus the dependence of local rearrangements on resolution.

6   Discussion

We have documented the behaviour of an estimator of the number of translocations intervening between two rearranged genomes, based only on the numbers of conserved syntenies on each chromosome, the lengths of the chromosomes and a simplified random model of interchromosomal exchange. In the absence of inversions, this estimator has undetectable bias up to 400 translocations, and has a rather moderate variance. Addition of high rates of inversion introduces some bias, though to detect this in our simulations we had to use many thousands of unrealistically large, hence maximally disruptive, inversions. If we also know the number of conserved segments, we can infer the number of inversions with corresponding accuracy.

The good properties of this estimator even after hundreds of translocations are remarkable given that it only explicitly takes into account the first-order effects of interchromosomal exchange. The fact that the introduction of chromosome sizes improves the estimator to such a degree compared to the previous version in [14], is somewhat surprising in that these sizes fluctuate greatly during the simulation.

Though this estimator is almost always well-defined and accurate (i.e., unbiased) for over 400 translocations in the human case, even in the presence of considerable intrachromosomal activity, we also explored a version of the estimator applicable to the situation with saturated chromosomes. This situation would be of some practical interest between 300 to 1000 translocations or more, though we are not aware of such cases being discussed in the biological literature. We found the estimator to be accurate after about 600 translocations, with a small but constant bias. We looked into this estimator because it was derived differently from eqn (3), but there are other, perhaps better, possibilities. For example, instead of solving eqn (3) for each chromosome separately, we are currently investigating the incorporation of eqn (4) into (3), so that we can solve

$$c^2 - \sum c^{(i)} = \sum_i \sum_{j \neq i} [1 - p_i(j)]^{2\hat{t}p(i)}, \tag{10}$$

for $\hat{t}$ directly for the entire genome, dispensing with the distinction between models for genomes having no saturated chromosomes and those having some.

In this paper, we applied our estimate to the human-mouse comparison at various levels of resolution. This showed that translocation estimates are extremely stable, while variability in the number of inversions inferred accounted for all the variation in the number of conserved segments due to differing levels of resolution. This reflects the discovery of high numbers of smaller-scale local arrangements recognizable from genomic sequence [6].

Our estimates of the number of translocations and inversions in the evolutionary divergence of man and mouse are only about a half of what has been published by Pevzner and Tesler [9, 10] who have attempted to reconstruct algorithmically the details of this history. Our model assumes each translocation and inversion creates two new segments, but the algorithms require a number of rearrangements almost equal to the number of segments to account for how the segments are ordered on the chromosomes. This accounts for the difference between the two sets of results. The reason the algorithms require one rearrangement per segment instead of one rearrangement per two segments is either

- Rearrangements almost always use at least one previously used breakpoint per rearrangement instead of two new ones, because breakpoints are largely confined to a small number of *fragile* regions on each chromosome, so that there is no parsimonious analysis of the segment orders which involves all or mostly two-breakpoint rearrangements, or
- Our two breakpoint per rearrangement model is correct, but the neglect, in the algorithmic approach, of segments smaller than a certain threshold value obscures the history and presents the algorithm with an effectively randomized order of segments along the chromosome [17]. Genomes with randomly ordered chromosomal segments tend to require one rearrangement per segment.

In any case, we also note that the proportion of inversions and translocations, if not their absolute numbers, is the same in our approach as in the results of the algorithmic approach.

Our model includes a feature that approximates the principle of conservation of the centromere. This principle prohibits translocations that result in one chromosome with no centromere and the other with two centromeres. In ongoing work we are attempting to drop this feature, since it is not always operative on the evolutionary time scale, taking into account such mechanisms as centromere inactivation and neocentromere activation, or chromosome fusion and chromosome fission.

Acknowledgements

Thanks to Aleksander Lenert and Phil Trinh for guidance with the genome browsers and other tools, to Adrian Maler for calculating the parameters of the inversion-size distribution, to David Kempe, Jon Kleinberg and David Liben-Nowell for discussions on the gossip problem, and to Nabil Benabbou for help with running our simulations on the High Performance Computing Virtual Laboratory facilities. Research supported by a Discovery grant to DS and and an undergraduate summer research scholarship to MM from the Natural Sciences and Engineering Research Council of Canada (NSERC). DS holds the Canada Research Chair in Mathematical Genomics and is a Fellow in the Evolutionary Biology Program of the Canadian Institute for Advanced Research.

References

1. Bed'hom, B. (2000). Evolution of karyotype organization in *Accipitridae*: A translocation model. In Sankoff, D. and Nadeau, J. H. (eds) *Comparative Genomics: Empirical and Analytical Approaches to Gene Order Dynamics, Map Alignment and Evolution of Gene Families*. Dordrecht, NL, Kluwer, 347–56.

2. Burt, D.W. (2002). Origin and evolution of avian microchromosomes *Cytogenetic and Genome Research*, **96**, 97–112.

3. De, A., Ferguson, M., Sindi, S. and Durrett, R. (2001). The equilibrium distribution for a generalized Sankoff-Ferretti model accurately predicts chromosome size distributions in a wide variety of species. *Journal of Applied Probability*, **38**, 324–34.

4. Hannenhalli, S. and Pevzner, P. A. (1995). Transforming men into mice (polynomial algorithm for genomic distance problem). *Proceedings of the IEEE 36th Annual Symposium on Foundations of Computer Science*. 581–92.

5. Housworth, E. A. and Postlethwait, J. (2002).Measures of synteny conservation between species pairs. *Genetics*, **162**, 441–8.

6. Kent, W. J., Baertsch, R., Hinrichs, A., Miller, W. and Haussler, D. (2003). Evolution's cauldron: Duplication, deletion, and rearrangement in the mouse and human genomes. *Proceedings of the National Academy of Sciences, USA*, **100**, 11484–9.

7. Mefford, H.C. and Trask, B.J. (2002). The complex structure and dynamic evolution of human subtelomeres. *Nature Reviews in Genetics*, **3**, 91-102; 229.

8. Nadeau, J. H. and Taylor, B. A. (1984). Lengths of chromosomal segments conserved since divergence of man and mouse. *Proceedings of the National Academy of Sciences, USA*,**81**, 814-8.

9. Pevzner, P. A. and Tesler, G. (2003). Genome rearrangements in mammalian genomes: Lessons from human and mouse genomic sequences. *Genome Research*, **13**, 37-45

10. Pevzner, P. A. and Tesler, G. (2003). Human and mouse genomic sequences reveal extensive breakpoint reuse in mammalian evolution. *Proceedings of the National Academy of Sciences, USA*, **100**, 7672-7

11. Sankoff, D., Deneault, M., Turbis, P. and Allen, C.P. (2002) Chromosomal distributions of breakpoints in cancer, infertility and evolution. *Theoretical Population Biology*, **61,** 497–501.
12. Sankoff, D. and Ferretti, V. (1996). Karotype distributions in a stochastic model of reciprocal translocation. *Genome Research*, **6,** 1–9.
13. Sankoff, D., Ferretti, V, and Nadeau, J.H. (1997) Conserved segment identification. *Journal of Computational Biology*, **4,** 559–65.
14. Sankoff, D., Parent, M.-N. and Bryant, D. (2000). Accuracy and robustness of analyses based on numbers of genes in observed segments. In Sankoff, D. and Nadeau, J. H. (eds) *Comparative Genomics: Empirical and Analytical Approaches to Gene Order Dynamics, Map Alignment and Evolution of Gene Families.* Dordrecht, NL, Kluwer, 299–306.
15. Sankoff, D., Parent, M.-N., Marchand, I. and Ferretti, V. (1997). On the Nadeau-Taylor theory of conserved chromosome segments. In Apostolico, A. and Hein, J. (eds) *Combinatorial Pattern Matching. Eighth Annual Symposium.* Lecture Notes in Computer Science **1264,** Springer Verlag, 262–74.
16. Sankoff, D. and Nadeau, J.H. (1996). Conserved synteny as a measure of genomic distance. *Discrete Applied Mathematics*, **71,** 247–57.
17. Sankoff, D. and Trinh, P. (2004). Chromosomal breakpoint re-use in the inference of genome sequence rearrangement. *Proceedings of RECOMB 04, Eighth International Conference on Computational Molecular Biology.* New York: ACM Press, 30–5.
18. Schubert, I. and Oud, J.L. (1997). There is an upper limit of chromosome size for normal development of an organism. *Cell*, **88,** 515–20.
19. Tesler, G. (2002). Efficient algorithms for multichromosomal genome rearrangements, *Journal of Computer and System Sciences*, **65,** 587–609.

Databases for Comparative Analysis
of Human-Mouse Orthologous Alternative Splicing

Bahar Taneri[*], Alexey Novoradovsky[*], Ben Snyder, and Terry Gaasterland

Laboratory of Computational Genomics
The Rockefeller University
1230 York Avenue
New York, NY 10021, USA
{bahar,anovo,ben,gaasterl}@genomes.rockefeller.edu

Abstract. Comparative analyses of alternative splicing across species can provide significant biological insight not only to evolution of alternative splicing,
but also to its regulation and functional significance. For comprehensive analyses of human and mouse alternatively spliced genes, we developed two databases of the human and the mouse transcriptomes, HumanSDB3 and MouSDB5
respectively. We showed that alternative splicing events in both of the transcriptomes are mainly due to the presence or absence of *internal cassette* exons.
Our databases allow in-depth analyses of alternative and constitutive exons
within alternatively spliced genes. Interactive web implementation of our databases brings to end-user the ability to instantly identify orthologous human-
mouse gene pairs with their corresponding exons. This is a novel visualization
method which provides easy access to conserved alternative splicing data and a
tool to explore the evolution of this important biological process.

1 Introduction

Alternative splicing is an important cellular process which generates several different
mRNA transcripts from a single gene, increasing the functional complexity of genomes [1,2]. This process enables production of structurally and functionally different proteins, expression of which can be tissue-specific, developmental-stage or
physiological-condition dependent [3]. More than 50% of the human genes are estimated to be alternatively spliced [4]. Given that alternative splicing occurs very frequently within genomes [5] and is conserved across species [6], it is of special interest to perform cross-species comparative analyses of alternative splicing.

Mouse and human genomes are highly conserved with about 80% of the mouse
genes having human orthologs. More than 90% of both the human and the mouse
genomes are shown to be within conserved syntenic regions [7]. Availability of extensive full-length transcript sequences and EST data both from the human genome
and from the mouse genome enables detailed analyses of alternative splicing within
and across these species.

In recent years, several different alternative splicing databases have been developed. Many of these have splice forms from one organism only, such as the Alterna-

---

[*] These authors have contributed equally to this work

J. Lagergren (Ed.): RECOMB 2004 Ws on Comparative Genomics, LNBI 3388, pp. 123–131, 2005.
© Springer-Verlag Berlin Heidelberg 2005

tive Splicing Annotation Project (ASAP) [8] and SpliceNest [9] human databases. Several databases contain alternative splice forms from two or more different organisms. These include Extended Alternatively Spliced EST Database (EASED) [10], the Putative Alternative Splicing database (PALSdb) [11], database of alternatively spliced genes (ASDB) [12], an alternative splice database of mammals (AsMamDB) [13] and database of canonical and non-canonical mammalian splice sites (SpliceDB) [14].

In a recent study, Kan *et al.* [15] employed cross-species analyses of human and mouse gene sets and identified novel alternative splice forms. Sugnet *et al.* [16] used whole genome alignments to assess conservation of alternative splicing. A subset of their human-mouse conserved splicing graphs is available at the UCSC genome browser. Some databases such as the Alternative Splicing Database (ASD) consortium are aiming to implement cross-species comparison features into their databases [17].

All of the existing alternative splicing databases and the cross-species comparison analyses are informative and facilitate further understanding of alternative splicing. However, to our knowledge, currently there is not a comparative alternative splicing database which allows instant human-mouse orthology comparison by the end-user for their gene of interest. With our web-implemented graphical user interface, we bring to the end-user the ability to find not only the orthologous alternatively spliced genes but also their corresponding alternative and constitutive exons.

In this paper, we describe two comprehensive databases of alternative splicing in human and mouse transcriptomes. We introduce the interactive web implementation of our databases which allows queries by the end-user to compare human and mouse alternatively spliced orthologs. Our study provides a novel visualization tool for conserved alternatively spliced genes and their corresponding alternative and constitutive exons.

2   Results

2.1   Definitions

For the cross-species comparative study described here, we developed two alternative splicing databases for the human and the mouse transcriptomes, HumanSDB3 [18] and MouSDB5 [19] respectively. These databases display the splice variants within the two genomes. They are developed as described by Taneri *et al.* [20] and for the work described here, we adopt Taneri *et al.*'s definitions and introduce new definitions of the following terms presented in italics.

Briefly, *transcripts* are either full-length mRNA or EST sequences. A *locus* is a genomic region which contains a set of overlapping transcripts called a *cluster*. A *cassette exon* is an alternative exon which is present in some transcripts and is absent from others. A *transcript-terminal cassette exon* is the 3' or 5' terminal exon within a transcript that matches no other internal exon in any other transcript. These exons map to genomic regions that are either introns or other transcript-terminal exons in other transcripts. A transcript-terminal cassette exon is only found in one transcript of the cluster. An *internal cassette exon* is a cassette exon which is not a terminal-transcript exon and is present in multiple transcripts within a cluster.

A *length-variant exon* is an exon with varying 5' or 3' splice sites or both. A cluster with one or more alternative exons is termed a *variant cluster*. All exons in an *invariant cluster* are constitutive with unique splice sites. A *genomic exon* is a stretch of uninterrupted nucleotides from the genomic DNA sequence each of which maps to a transcript within a cluster.

2.2 HumanSDB3 – MouSDB5 Cluster, Transcript and Alternative Exon Analyses

Our analyses revealed that HumanSDB3 contains a total of 20707 clusters, 81.3% of which are variant. 74% of the 20090 MouSDB5 clusters are variant (Table 1). These results indicate that majority of the loci within both of these genomes are variant and thus produce several alternatively spliced transcripts (Fig. 1). We observed that there is a higher variation in the human transcriptome compared to the mouse transcriptome (Fig. 1). This phenomenon might be explained by the fact that there were 1313307 more human input transcripts. Thus the final mapped human transcripts were higher in number compared to the final number of mouse transcripts (Table 1). As a result, average number of transcripts per HumanSDB3 cluster is 70.5 whereas average number of transcripts per MouSDB5 cluster is 57.2. Consequently, total number of human exons are higher than the total number of mouse exons and there are more exons per HumanSDB3 cluster than exons per MouSDB5 cluster (Table 1).

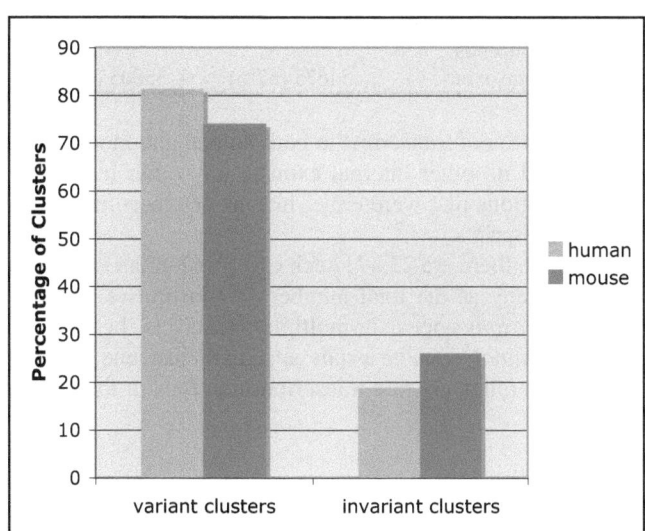

Fig. 1. Distribution of variant and invariant clusters in HumanSDB3 and MouSDB5. *Blue bars* show HumanSDB3 cluster counts, *red bars* show MouSDB5 cluster counts

Table 2 shows that alternative splicing both in human and in mouse transcriptomes is mainly due to presence or absence of *internal cassette exons*. These exons make up 67% and 62% of the total number of alternative exons in human and in mouse transcriptomes respectively.

Table 1. Statistical Analysis of HumanSDB3 and MouSDB5

	HumanSDB3	MouSDB5	Mouse as % of human
Total number of clusters	20707	20090	97%
Number of invariant clusters	3881	5221	134%
Number of variant clusters	16826	14869	88%
Number of input transcripts	4635471	3322164	71%
Number of mapped transcripts	1459966	1149658	78%
% of input transcripts mapped	31%	35%	--
Average transcripts per cluster	70.5	57.2	81%
Total number of exons	241824	216432	89%
Average number of exons per cluster	11.7	10.8	92%

Table 2. Alternative Exon Analysis of HumanSDB3 and MouSDB5

	HumanSDB3	MouSDB5	Mouse as % of human
Total number of alternative exons	81203	56494	70%
Number of constitutive, length-variant exons with both 5' and 3' variable ends	1978 (2%)	1140 (2%)	58%
Number of constitutive, length-variant exons with only 5' variable ends	12761 (16%)	10721 (19%)	84%
Number of constitutive, length-variant exons with only 3' variable ends	11789 (15%)	9628 (17%)	81%
Number of internal-cassette exons	54675 (67%)	35005 (62%)	64%

A substantial proportion of transcripts in both human and mouse had 3' or 5' terminal exons that matched no other internal exon in any other transcript. Rather, they mapped to genomic regions that were either introns or other *transcript-terminal cassette exons* in other transcripts.

As shown in Table 3, there are 23,471 such exons in human and 16,158 such exons in mouse, nearly as many as the total number of constitutive length-variant exons. Most *internal cassette exons* appear in multiple transcripts. In contrast, the majority of these *transcript-terminal cassette exons* appear in just one transcript, indicating that they may be due to relatively rare transcription events, or that they may be due to cloning artifacts.

Table 3. Transcript-terminal cassette exons in HumanSDB3 and MouSDB5

	HumanSDB3	MouSDB5	Mouse as % of human
Total number of alternative exons (including transcript-terminal exons)	104674	72652	69%
Number of transcript-terminal cassette exons	23471 (22%)	16158 (22%)	69%
Total number of constitutive length-variant exons	26528 (25%)	21489 (30%)	81%
Number of internal cassette exons	54675 (52%)	35005 (48%)	64%

2.3   Web Implementation and Comparative Species Feature of HumanSDB3 and MouSDB5

Our databases are implemented with interactive graphical user interfaces and can be browsed by users. HumanSDB3 is available at http://genomes.rockefeller.edu/autodb/ sdb.php?db=HumanSDB3 and MouSDB5 is available at http://genomes.rockefeller. edu/autodb/sdb.php?db=MouSDB5. Using these interfaces users can access database statistics on clones, chromosomes, clusters and exons. These databases can be browsed for any of the splice clusters and information on transcripts mapping to those clusters are provided. When available, annotation and library information for each transcript together with the nucleotide sequences of their exons are given. Splice clusters can be accessed by providing as input any of the following parameters of the user's gene of interest: annotation, chromosome number, cluster ID, cluster type or UniGene ID.

Figure 2 shows a variant human cluster which is annotated as the General Transcription Factor IIH, polypeptide 2. This figure displays a partial view of the full-length transcripts and the ESTs which map to this variant loci on human chromosome 5. The complete cluster can be viewed at http://genomes.rockefeller.edu/autodb/ cluster_map.php?cluster_id=Hs.3.chr5n.15596&db=HumanSDB3.

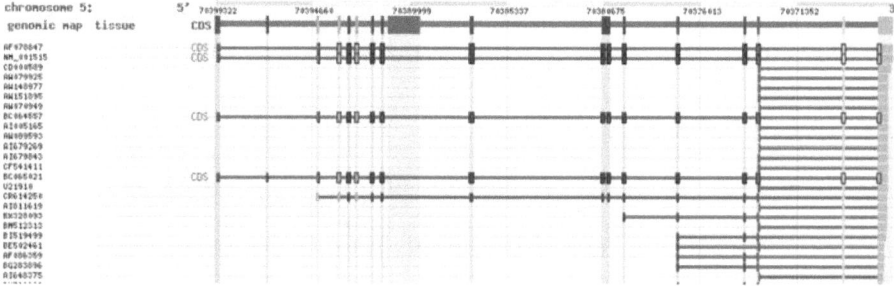

Fig. 2. Human General Transcription Factor IIH, polypeptide 2. Screenshot from HumanSDB3 cluster Hs.3.chr5n.15596. This figure displays a partial view of the ESTs and full-length transcripts mapping to this variant loci on human chromosome 5. The very first line above the transcripts is the *genomic exon map* of this variant cluster (labeled as *genomic map* on upper left corner)

Figure 3 shows the partial view of the variant cluster annotated as the mouse General Transcription Factor IIH, polypeptide 2. The complete cluster along with further information of transcripts and exons can be found at http://genomes.rockefeller.edu/ autodb/cluster_map.php?cluster_id=Mm.5.chr13n.4377&db=MouSDB5.

Our *compare species feature* allows the end-user to compare these two orthologous transcripts and their corresponding exons. As shown in Figure 4, our program aligns the human and mouse genomic exon maps. Lines going across the opposite exons indicate the orthologous human-mouse exons. The *compare species feature* of our databases can be accessed at http://genomes.rockefeller.edu/autodb/compare_ 1.php? db1=HumanSDB3&db2=MouSDB5. The end-user is able to find their gene-pair of interest by providing as input any of the following parameters: Keyword, Gene Symbol, Splicing Cluster ID, GenBank accession number or UniGene cluster ID.

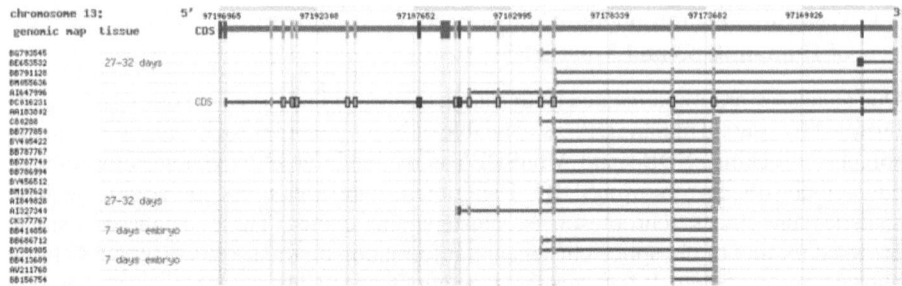

Fig. 3. Mouse General Transcription Factor IIH, polypeptide 2. Screenshot from MouSDB5 cluster Mm.5.chr13n.4377. This figure displays a partial view of the ESTs and full-length transcripts mapping to this variant loci on mouse chromosome 13. The very first line above the transcripts is the *genomic exon map* of this variant cluster (labeled as *genomic map* on upper left corner)

Database of Splicing Variants. Comparison of Hs.3.chr5n.15596 and Mm.5.chr13n.4377.

Hs.191356: General transcription factor IIH, polypeptide 2, 44kDa

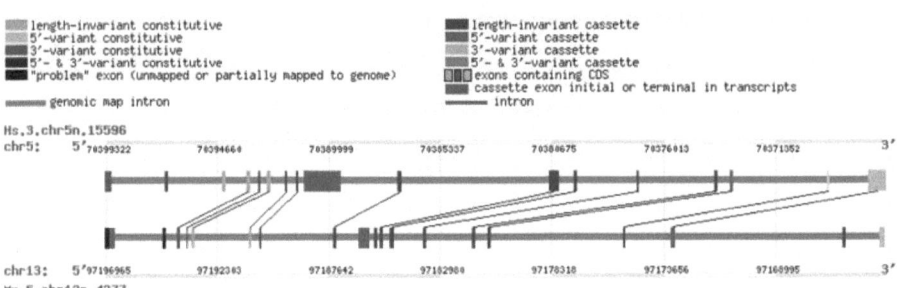

Fig. 4. Comparison of exons in an orthologous human – mouse gene pair. Orthologous transcripts of General Transcription Factor IIH, polypeptide 2. *Upper genomic map* is from HumanSDB3 variant cluster Hs.3.chr5n.15596 and *lower genomic map* is from MouSDB5 variant cluster Mm.5.chr13n.4377. Lines connecting opposite exons indicate their orthology

3   Materials and Methods

3.1   Development of HumanSDB3 and MouSDB5

The HumanSDB3 and MouSDB5 databases have been developed using the methods described by Taneri *et al.* [20] and the references therein. Input transcript sequences were downloaded from UniGene human version no. 173 and mouse version no. 139. There were a total of 4635471 human and a total of 3322164 mouse input transcript sequences. These transcripts sequences were aligned to the genome using blat [21]. Top 10% matches from blat reports were aligned to the genomic region by SIM4 [22] and the top scoring match was considered to be the best alignment. The following criteria were applied to each best-aligned transcript sequence for inclusion in the final database. The transcript had 75% or greater identity to the genome. In addition, every exon of the transcript had either a 95% identity to the genome or contained 5 or less mismatches. Each transcript had a minimum of two exons. Final versions of the databases contained a total of 1459966 human transcripts and a total of 1149658 mouse transcripts. Final clusters had a minimum of three transcripts each.

3.2 Computation of Orthologous Splicing Clusters

Computation of orthologous splicing clusters for the *compare species feature* of HumanSDB3 and MouSDB5 was carried out as follows. FASTA files were created out of spliced genomic exons of each splice cluster in HumanSDB3 and in MouSDB5. These files were blasted against each other with blastall options –p blastn –m 8 –e 1e-20 [23]. The corresponding pair-wise matches were considered as potentially orthologous. Coordinates of syntenic chromosome regions for human and mouse were downloaded from http://nbcr.sdsc.edu/GRIMM/HMR_Aug2003/blocks_hmr_ 1000000. Chromosome coordinates of potentially orthologous pairs were compared with the coordinates of syntenic regions. Orthologous pairs were established when the splicing cluster coordinates fell within the syntenic regions.

4 Discussion

Our analyses of the human and mouse transcriptomes show that alternative splicing is widespread within both species. In both of the transcriptomes alternative splicing is mainly due to the presence or absence of *internal cassette exons*. This finding indicates the functional significance of this type of alternative exons. We also observe the high prevalence of *transcript-terminal cassette exons* in both of the transcriptomes. These exons do not match any other internal exons. They map to genomic regions which are either intronic or other transcript-terminal exons. These exons appear only in one transcript. Whereas *internal-cassette exons* appear in multiple transcripts. Based on these results, we conclude that *transcript-terminal exons* might be either due to relatively rare transcription events or due to cloning artifacts.

We observe that variation in human transcriptome is higher than the variation in mouse transcriptome. This phenomenon might be explained by the fact that there were about a million more human input transcripts than mouse input transcripts. Consequently, more human transcripts entered the final database. Thus there are more transcripts per cluster in HumanSDB3 than in MouSDB5 and the total number of exons in human is higher than it is in mouse.

In this study, we introduce a novel web-based visualization method to study alternative splicing. This tool brings to the end-user the ability to analyze alternative splicing in their gene of interest. Users can view all exons of their gene, access their nucleotide sequences and learn about the libraries of the transcripts sequenced for that gene. In addition, users are able to find orthologous pairs for their genes of interest and to study pairs of orthologous human-mouse constitutive and alternative exons in detail.

The work described here will have significant implications in further understanding the evolution of alternative splicing. We provide a unique method for studying conservation of alternative splicing. Our databases expand the knowledge on human-mouse transcriptome conservation. Our visualization method for orthologous exon pairs provides a means for comparative studies of these two species.

Easy access to comparative alternative splicing data through these databases and the visualization tool allows instant information retrieval about variant exons and their sequences. This in turn further aids in experimental design for alternative splicing studies.

Acknowledgements

We acknowledge support from Mathers Foundation and Hirschl Foundation. This work has been partially funded by NSF grant DBI9984882 and NIH grant GM62529 to T.G. We thank Anna Neill for useful discussions and the members of Laboratory of Computational Genomics for their support. Corresponding author T.G. can be reached at gaasterl@genomes.rockefeller.edu as well as at gaasterland@ucsd.edu.

References

1. Black, D.L.: Protein diversity from alternative splicing: a challenge for bioinformatics and post-genome biology. Cell 103 (2000) 367-370
2. Brett, D., Popisil, H., Valcarel, J., Reich, J., Bork, P.: Alternative splicing and genome complexity. Nat Genet. 1 (2002) 29-30
3. Graveley, B.R.: Alternative splicing: increasing diversity in the proteomic world. Trends Genet. 17 (2001) 100-107
4. Modrek, B., Lee, C.: A genomic view of alternative splicing. Nature Genet. 30 (2002) 13-19
5. Modrek, B., Resch, A., Grasso, C., Lee, C.: Genome-wide detection of alternative splicing in expressed sequences of human genes. Nucleic Acids Res. 29 (2001) 2850-2859
6. Thanaraj, T.A., Clark, F., Muilu, J.: Conservation of human alternative splice events in mouse. Nucleic Acids Res. 31 (2003) 2544-2552
7. Mouse Genome Sequencing Consortium: Initial sequencing and comparative analysis of the mouse genome. Nature. 420 (2002) 520-562
8. Lee, C., Atanelov, L., Modrek, B., Xing, Y.: ASAP: the Alternative Splicing Annotation Project. Nucleic Acids Res. 31 (2003) 101-105
9. Coward, E., Haas, S.A., Vingron, M.: SpliceNest: visualizing gene structure and alternative splicing based on EST clusters. Trends Genet. 18 (2002) 53-55
10. Pospisil, H., Herrmann, A., Bortfeldt, R.H., Reich, J.G.: EASED: Extended Alternatively Spliced EST Database. Nucleic Acids Res. 32 (2004) D70-74
11. Huang, Y-H., Chen, Y-T., Lai, J-J., Yang, S-T., Yang, U-C.: PALS db: Putative Alternative Splicing database. Nucleic Acids Res. 30 (2002) 186-190
12. Dralyuk, I., Brudno, M., Gelfand, M.S., Zorn, M., Dubchack, I.: ASDB: database of alternatively spliced genes. Nucleic Acids Res. 28 (2000) 296-297
13. Ji, H., Zhou, Q., Wen, F., Xia, H., Lu, X., Li, Y.: AsMamDB: an alternative splice database of mammals. Nucleic Acids Res. 29 (2001) 260-263
14. Burset, M., Seledtsov, I.A., Solovyev, V.V.: SpliceDB: database of canonical and non-canonical mammalian splice sites. Nucleic Acids Res. 29 (2001) 255-259
15. Kan, Z., Castle, J., Johnson, J.M., Tsinoremas, N.F.: Detection of novel splice forms in human and mouse using cross-species approach. Pac. Symp. Biocomput. (2004) 42-53

16. Sugnet, C.W., Kent, W.J., Ares, M. Jr, Haussler, D.: Transcriptome and genome conserva-
    tion of alternative splicing events in humans and mice. Pac. Symp. Biocomput. (2004) 66-
    77
17. Thanaraj, T.A., Stamm, S., Clark, F., Riethoven, J.J., Le Texier, V., Muilu, J.: ASD: the
    Alternative Splicing Database. Nucleic Acids Res. 32 (2004) D64-D69
18. HumanSDB3 [http://genomes.rockefeller.edu/autodb/sdb.php?db=HumanSDB3]
19. MouSDB5 [http://genomes.rockefeller.edu/autodb/sdb.php?db=MouSDB5]
20. Taneri, B., Snyder, B., Novoradovsky, A., Gaasterland, T.: Alternative splicing of mouse
    transcription factors affect their DNA-binding domain architecture and is tissue specific.
    Genome Biology. 5 (2004) R75
21. Kent, W.J.: BLAT—the BLAST-like alignment tool. Genome Res. 12 (2002) 656-664
22. Florea, L., Hartzell, G., Zhang, Z., Rubin, G.M., Miller, W.: A computer program for
    aligning a cDNA sequence with a genomic DNA sequence. Genome Res. 8 (1998) 967-
    974
23. Altschul, S.F., Gish, W., Miller, W., Myers, E.W., Lipman, D.J.: Basic local alignment
    search tool. J Mol Biol. 215 (1990) 403-10

Author Index

Lecture Notes in Bioinformatics